SpringerBriefs in Applied Sciences and Technology

SpringerBriefs present concise summaries of cutting-edge research and practical applications across a wide spectrum of fields. Featuring compact volumes of 50 to 125 pages, the series covers a range of content from professional to academic.

Typical publications can be:

- A timely report of state-of-the art methods
- An introduction to or a manual for the application of mathematical or computer techniques
- A bridge between new research results, as published in journal articles
- A snapshot of a hot or emerging topic
- An in-depth case study
- A presentation of core concepts that students must understand in order to make independent contributions

SpringerBriefs are characterized by fast, global electronic dissemination, standard publishing contracts, standardized manuscript preparation and formatting guidelines, and expedited production schedules.

On the one hand, **SpringerBriefs in Applied Sciences and Technology** are devoted to the publication of fundamentals and applications within the different classical engineering disciplines as well as in interdisciplinary fields that recently emerged between these areas. On the other hand, as the boundary separating fundamental research and applied technology is more and more dissolving, this series is particularly open to trans-disciplinary topics between fundamental science and engineering.

Indexed by EI-Compendex, SCOPUS and Springerlink.

More information about this series at http://www.springer.com/series/8884

Kok Keong Choong · Jayaprakash Jaganathan ·
Sharifah Salwa Mohd Zuki · Shahiron Shahidan ·
Nurul Izzati Raihan Ramzi Hannan

Concrete-Filled Double Skin Steel Tubular Column with Hybrid Fibre Reinforced Polymer

Post Fire Repair

 Springer

Kok Keong Choong
School of Civil Engineering
Universiti Sains Malaysia
Penang, Malaysia

Jayaprakash Jaganathan
School of Civil Engineering
Vellore Institute of Technology
Vellore, Tamil Nadu, India

Sharifah Salwa Mohd Zuki
Department of Civil Engineering, Faculty
of Civil Engineering and Built Environment
Universiti Tun Hussein Onn
Johor, Malaysia

Shahiron Shahidan
Department of Civil Engineering, Faculty
of Civil Engineering and Built Environment
Universiti Tun Hussein Onn Malaysia
Johor, Malaysia

Nurul Izzati Raihan Ramzi Hannan
Department of Civil Engineering, Faculty
of Civil Engineering and Built Environment
Universiti Tun Hussein Onn Malaysia
Johor, Malaysia

ISSN 2191-530X   ISSN 2191-5318 (electronic)
SpringerBriefs in Applied Sciences and Technology
ISBN 978-981-16-2714-9   ISBN 978-981-16-2715-6 (eBook)
https://doi.org/10.1007/978-981-16-2715-6

This Springer imprint is published by the registered company Springer Nature Singapore Pte Ltd.
The registered company address is: 152 Beach Road, #21-01/04 Gateway East, Singapore 189721, Singapore

Foreword

Fire is one of the worst conditions that a structure may be subjected to. Hence, behavior of the structure and material that underwent fire exposure needs to be clearly understood so that the engineers can come up with the most effective solution to deal with the consequences. Most of the time, structure that was exposed to fire will be demolished. However, researches show that factors such as the maximum temperature of the fire, the duration of the fire and even the method of distinguishing the fire play an important role in post-fire performance of the structures. Taking all this into consideration, the structures might not need to be demolished. Hence, repairing the damaged structures is the most economically and environmentally sustainable option.

This book provides useful information on the post-fire behavior of concrete-filled double skin steel tubular (CFDST) columns. In addition, the fire-damaged CFDST columns were repaired using hybrid fiber-reinforced polymer (FRP). Hopefully, this book will give an insight into both the behavior and performance of the CFDST columns after fire exposure and after being repaired with hybrid FRP. Thus, more researchers and engineers can benefit from this knowledge.

Sharifah Salwa Mohd Zuki
Department of Civil Engineering
Universiti Tun Hussein Onn Malaysia
Johor, Malaysia

Preface

The concrete-filled double skin steel tubular (CFDST) column is becoming more popular nowadays due to its superior performance compared to conventional composite column and concrete-filled steel tubular (CFST) column. However, the use of this type of column is still limited to outdoor construction such as bridge piers and transmission towers where fire is not the main concern. Moreover, existing research studies on the CFDST column only focused on fire performance, and limited research studies can be found on the residual strength of the CFDST column. Residual strength can be used to determine the most suitable repair method needed in order to retrofit the column. Therefore, this study aims to study the effect of different parameters on the residual strength of the CFDST column. Among discussed parameters are the thickness of the outer steel tube (t_0) and fire exposure time. In addition, this study also aims to determine the effectiveness of the repair method using Single and Hybrid fiber-reinforced polymer (FRP) of fire-damaged CFDST columns. CFDST columns were heated in accordance with ASTM E119-11: Standard Test Methods for Fire Tests of Building Construction and Materials until the temperature reached 600 °C. Afterwards, the temperature was kept constant for two different durations, i.e., 60 and 90 mins. The specimen was then left to cool down to room temperature inside the furnace before it was taken out and repaired by Single and Hybrid FRP. The specimens were categorized into the following three groups: (1) unheated or control specimens, (2) heated and unrepaired specimens and (3) heated and repaired specimens. All specimens were subjected to axial compression loading until failure. The first and second category specimens failed by local outward buckling of outer steel tube, crushing of concrete and local buckling of inner steel tube, whereas specimens in the third category failed by rupture of FRP followed by similar local buckling and concrete crushing as those observed in first and second category specimens. Ultimate strength, secant stiffness and Ductility Index (DI) decreased as the temperature of the specimen increased. The loss in secant stiffness of thinner CFDST specimens exposed to 60 mins of fire exposure time is similar to thicker CFDST specimens exposed to 90 mins of fire exposure time regardless of their diameter. In addition, CFDST specimens exposed to 90 mins of fire exposure time were more ductile than control specimen. RSI and secant stiffness increased with the increase in fire exposure time. Interestingly, the highest RSI achieved is only 22% which means the specimens were still able to carry more than 70% of their initial load after being exposed to 90 mins of fire exposure time with only 3 mm thickness of outer steel tube. Repairing the fire-damaged CFDST columns with Single and Hybrid FRP is proven to improve the ultimate compressive strength significantly. The increment in ultimate compressive strength is more pronounced in the specimen with Hybrid FRP and thinner outer steel tube. The secant stiffness and Ductility Index (DI) of repaired specimens were, however, not able to be restored to those of the control specimen.

Penang, Malaysia
Vellore, India
Johor, Malaysia
Johor, Malaysia
Johor, Malaysia

Kok Keong Choong
Jayaprakash Jaganathan
Sharifah Salwa Mohd Zuki
Shahiron Shahidan
Nurul Izzati Raihan Ramzi Hannan

Acknowledgements

Alhamdulillah, all praises to Allah SWT, the Al-Mighty, for blessing me with strength and motivation to complete this research. I would like to thank my parents (Mohd Zuki Yusof and Kalsom Md. Saad), my parents-in-law (Dah Osman and Abd. Rahim Jusoh), my husband (Shahiron Shahidan) and my daughter (Zara Amani) for their external support, love and encouragement throughout my studies. They have made me more confident and make the obstacles more bearable.

Special thanks and my deepest appreciation to my supervisor Prof. Ir. Dr. Choong Kok Keong. He was very supportive and more than willing to share his time and knowledge with me without hesitation. I would like to thank my co-supervisor Prof. Dr. J. Jayaprakash for giving me his utmost help and advice.

I am deeply grateful to Jamilus Research Center, Faculty of Civil Engineering and Built Environment, Universiti Tun Hussein Onn Malaysia (UTHM), for enabling me to publish this book. I would like to express my gratitude to the many people who saw me through this book: those who provided support, talked things over, read, wrote, offered, commented and allowed me to quote their remarks and assisted in the editing, proofreading and design this book. Last but not least, I beg for forgiveness from all people who have been with me during the editing process of this book, whose names I have failed to mention.

November 2020   Sharifah Salwa Mohd Zuki

Contents

Abbreviations

CFDST	Concrete-Filled Double Skin Steel Tubular
CFRP	Carbon Fiber-Reinforced Polymer
CFST	Concrete-Filled Steel Tubular
DI	Ductility Index
FRP	Fiber-Reinforced Polymer
GFRP	Glass Fiber-Reinforced Polymer
HSC	High Strength Concrete
HSS	Hollow Structural Section
NSC	Normal Strength Concrete
RSI	Residual Strength Index
SEI	Strength Enhancement Index

Symbols

$\Delta_{0.85}$ Displacement at 0.85 ultimate strength
Δ_u Displacement at ultimate strength
N_{eS} Ultimate strength of repaired CFDST columns
N_{eU} Ultimate strength of unrepaired CFDST columns
N_u Ultimate strength of CFDST column at ambient temperature
$N_u(t)$ Ultimate strength of CFDST column after fire
t_0 Thickness of outer steel tube

Chapter 1
Introduction

Steel hollow structural section (HSS) are widely used in high rise building and as bridge pier because of their resistance to lateral movement in addition to its lighter weight compared with solid steel section and reinforced concrete columns [1]. HSS columns are also known to be very effective in resisting compression loads and are widely used especially in industrial building as framed structures [2]. Filling this hollow column with plain concrete leads to a number of benefits such as increasing the load bearing capacity of the columns, higher fire resistance compared with HSS without concrete filling, preventing spalling of concrete when subjected to fire due to existence of steel and finally, the presence of steel eliminates the need of formwork [3–5] thus, leading to a rapid [6, 7] and economical construction [8]. Over time, engineers began to use concrete-filled hollow steel column or also known as concrete-filled steel tubular (CFST) column to replace HSS due to the above mentioned advantages. Overalls, CFST column are proven to be more economical than HSS [1].

The profile of concrete-filled double skin steel tubular (CFDST) column is similar to CFST except for the void in the middle of the column as shown in Fig. 1.1. CFDST columns have been used bridge piers in Japan, owing to its good damping and energy absorption properties as well as light weight cross-section [9]. More recently, Han et al. [10] reported that CFDST columns have been used as an electric pole in China (Fig. 1.2). Unlike CFDST columns, CFST columns have been widely used in China for almost 50 years. Among the examples are (1) Ruifeng building in Hangzhou, (2) Zhaohua Jialing River Bridge and (3) Qianmen subway station in Beijing [10]. Furthermore, CFDST columns are used only in outdoor construction where fire is not the main concern.

K. K. Choong et al., *Concrete-Filled Double Skin Steel Tubular Column with Hybrid Fibre Reinforced Polymer*, SpringerBriefs in Applied Sciences and Technology, https://doi.org/10.1007/978-981-16-2715-6_1

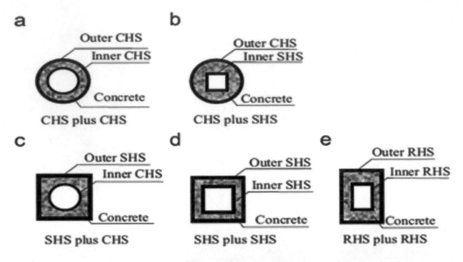

a Outer CHS / Inner CHS / Concrete — CHS plus CHS

b Outer CHS / Inner SHS / Concrete — CHS plus SHS

c Outer SHS / Inner CHS / Concrete — SHS plus CHS

d Outer SHS / Inner SHS / Concrete — SHS plus SHS

e Outer RHS / Inner RHS / Concrete — RHS plus RHS

Fig. 1.1 Typical profile of concrete-filled double skin tubular column [11]

Fig. 1.2 A CFDST pole in China [10]

References

1. Lam, D., Williams, C.A.: Experimental study on concrete filled square hollow sections. Steel Compos. Struct. **4**(2), 95–112 (2004)
2. Kodur, V.K.R., Lie, T.T.: Experimental Studies on the Fire Resistance of Circular Hollow Steel Columns Filled with Steel-Fibre-Reinforced Concrete, p. 1995. National Research Council Canada, Institute for Research and Construction (1995)
3. Han, L.H., Yang, H., Cheng, S.L.: Residual strength of concrete filled RHS stub columns after exposure to high temperatures. Adv. Struct. Eng. **5**(2), 123–134 (2002)
4. Han, L.H., Zhao, X.L., et al.: Experimental study and calculation of fire resistance of concrete-filled hollow steel columns. J. Struct. Eng. **129**(3), 346–356 (2003)
5. Han, L.H., Xu, L., Zhao, X.L.: Tests and analysis on the temperature field within concrete filled steel tubes with or without protection subjected to a standard fire. Adv. Struct. Eng. **6**(2), 121–133 (2003)
6. Han, L.., Huo, J.S. Wang, Y.C.: Compressive and flexural behaviour of concrete filled steel tubes after exposure to standard fire. J. Constr. Steel Res. **61**(7), 882–901 (2005)
7. Yang, Y.F., Han, L.H.: Concrete-filled double-skin tubular columns under fire. Mag. Concr. Res. **60**(3), 211–222 (2008)
8. Tao, Z., Han, L.H., Wang, L.L.: Compressive and flexural behaviour of CFRP-repaired concrete-filled steel tubes after exposure to fire. J. Constr. Steel Res. **63**(8), 1116–1126 (2007)
9. Zhao, X.L., Han, B., Grzebieta, R.H.: Plastic mechanism analysis of concrete-filled double-skin (SHS inner and SHS outer) stub columns. Thin-Walled Struct. **40**(10), 815–833 (2002)
10. Han, L.H., Li, W., Bjorhovde, R.: Developments and advanced applications of concrete-filled steel tubular (CFST) structures: members. J. Constr. Steel Res. **100**, 211–228 (2014)
11. Lu, H., Han, L.H., Zhao, X.L.: Fire performance of self-consolidating concrete filled double skin steel tubular columns: experiments. Fire Saf. J. **45**(2), 106–115 (2010)

Chapter 2
Literature Review

CFDST columns were first introduced by Shakir-Khalil in 1990s and many researchers studied CFDST columns since then [1]. The CFDST columns introduced by Shakir-Khalil at that time have very large void ratio with only 12 mm spacing between the inner and outer tube. The spacing between these two tubes was filled with 2 types of filler materials, grout and micro-concrete. Shakir-Khali, suggested that, in order to increase the behavior of CFDST column, the spacing between the tubes need to be increased, which allows normal density concrete to be used as filler materials instead of grout and micro-concrete. However, Tao et al. concluded that the existence of void ratio did not affect the performance of CFDST columns [2].

However, many research studies only focused on behavior of CFDST at ambient temperature such as [2–7] and [8]. Through extensive literature review, there are only five papers founds (three on finite element modeling) which focused on fire performance of CFDST columns [9–13]. Therefore, post fire repair of CFST columns is taken as reference when discussing this topic.

This chapter describes previous research studies on CFST and CFDST columns exposed to fire and post fire repair of CFDST columns with fiber reinforced polymer (FRP). Since there are limited number of research studies on CFDST columns exposed to fire and post fire repair of CFDST columns, research studies on concrete-filled steel tubular (CFST) columns are also included in this literature review.

2.1 Concrete-Filled Double Skin Steel Tubular (CFDST) Columns

CFDST columns possess similar benefits as CFST column. However, the presence of inner steel tube gave CFDST columns additional benefits compared to CFST. On the other hand, it also distinguishes CFDST from CFST. Both outer and inner steel

© The Author(s), under exclusive license to Springer Nature Singapore Pte Ltd. 2021
K. K. Choong et al., *Concrete-Filled Double Skin Steel Tubular Column with Hybrid Fibre Reinforced Polymer*, SpringerBriefs in Applied Sciences and Technology,
https://doi.org/10.1007/978-981-16-2715-6_2

tube provides confinement to the concrete. This will increase the column strength and ductility and as for steel, the concrete will reduce the possibility of local buckling for both inner and outer tube [14–16].

In addition to that, CFDST column when compared with conventional reinforced concrete possesses smaller cross section under similar load. Therefore, this will increase usable space especially for a structure that requires larger spaces to function such as office building and parking lots [17]. All the above advantages of concrete-filled hollow section are contributed by composite action of concrete and steel. Concrete core acts as component to resist compressive forces and at the same time preventing or suppressing earlier buckling of steel members whereas, steel elements not only reinforces the concrete but also resist any tensile forces, bending moment as well as shear forces [18]. For CFDST or CFST column without fire protection, the elimination of fire protection will increase its aesthetic value, reduce additional cost to provide fire protection as well as maintenance of fire protection. It can also increase usable spaces in building [19].

The advantages of CFDST columns are similar to those of CFST columns with a few additional benefits due to its unique characteristics. The summary of advantages possesses by CFDST columns are as follows:

(a)   Substantial increase in load bearing capacity of columns compared with traditional composite columns [20].

(b)   Higher resistance to fire even without fire protection layers for the outer steel tube [20].

(c)   Elimination of the need of formwork due to outer steel thus speed up construction processes and allow work to be done in all types of weather [20].

(d)   Prevention of spalling of concrete in the event of fire by outer steel tube [21].

(e)   Improvement in column stability by filling the HSS with concrete [22].

(f)   Increase in strength, ductility and energy absorption of the HSS columns by filling the steel columns with concrete [22].

(g)   Aesthetic especially when the steel used is an unprotected steel column [23].

(h)   Reduction in size of column footing [23].

(i)   Increase in bending stiffness of the columns [22].

(j)   Prevention of instability under external pressure [22].

(k)   Increase in local stability due to composite action between all three components [22].

(l)   Increase in global stability with the increased section modulus [22].

(m) Better seismic resistance [22].

(n)   Lightweight [22].

(o)   Good damping characteristics [22].

(p)   Good in resisting cyclic loading [22].

Due to its benefits over conventional reinforced concrete column and CFST column, CFDST was said to have potential to be used as columns in high rise building or pier in structures [6, 11]. CFDST can also be used as compression chambers in deep-sea pressure vessel, in the legs of offshore platforms in deep water, to large diameter columns and to structures subjected to ice loading [3, 4]. In Japan, CFDST have been used as high rise bridge piers in order to reduce weight of structure. Furthermore, the piers still maintain a large energy absorption capacity against earthquake loading despite the reduction in weight [3, 4].

2.2 Background on Fire Test of Structural Elements

Fire safety is an important feature in designing structures so that the structures will maintain their integrity during fire. In other words, the structures need to withstand partial or total collapse and at the same time preventing the spread of fire as well as loss of life for a certain period [24]. Different elements of structures (i.e., beam, column) have different provision of fire resistance. They vary from building to building depending on the quantity of combustible present (the fire load), the amount of ventilation provided by the design and the geometry of the floor where the fire occurs [24].

In the event of fire, fire safety is first addressed by the provision of active fire protection system such as, sprinkler systems. Active fire protections are meant to delay or if possible to stop the fire from spreading and becoming fully developed. If the fire becomes fully developed and active fire protection system fails to prevent this, fire safety then shifts towards ensuring the structural integrity of the structure for certain period of times so that the occupant can evacuate or until the arrival of fire fighters. During this period of time, structures have to possess ability to prevent the spread of fire and collapse [25]. The integrity of the structures can be ensured by providing adequate cover to reinforcement, adequate fire protection layers, compartmentalization, etc. In term of fire safety, this is called passive fire protections [26] and this is shown in Fig. 2.1.

2.2.1 Fire Test

In order to identify the fire resistance of a structure, researchers run fire test on building elements separately, such as beams, columns, floors and walls. Individual element tests do not indicate the exact performance of the whole structure but the results and observations can be used as a representative of a whole structure test since fire tests on structures or buildings are rarely conducted due to financial constraint and the complexity of the structures itself.

Real fire test is rarely done because it is costly and it takes longer time to set up and run the test [27], unlike test using standard fire test. However, standard fire test does not portray the real condition and scenario during fire. Hence, effort has been made so that the standard fire test follows as close as possible stages occurring in real fire. Even though there are major differences between real fire and standard fire,

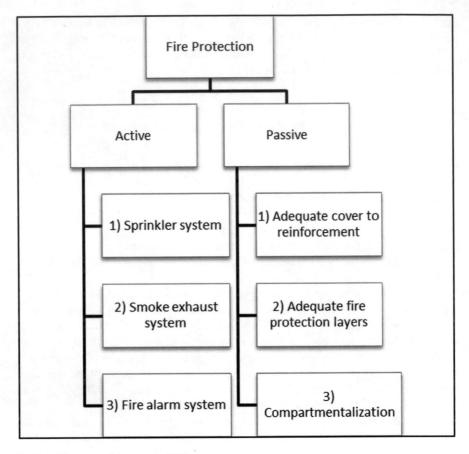

Fig. 2.1 Fire protection systems [26]

researchers can still use the data with confidence from educating themselves with the limitation and way to interpret the data from standard fire test [28].

There are three stages of fire that standard fire curve has to follow in order to portray real fire situation, namely; (1) fire growth; (2) fully developed; and (3) decay [29]. The time-temperature curve mention earlier is shown in Fig. 2.2.

2.2.2 Fire Test in Standards and Codes

There are a few design codes for structural element test that can be used to test fire performance as shown in Table 2.1. Generally, different country has different design codes. However, design codes such as ASTM E119 (2010), ISO-834 (2014) and BS 476 (1987) are widely used around the world.

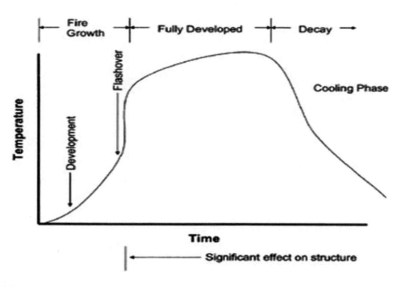

Fig. 2.2 Time-temperature curve of fire [26]

Table 2.1 Design codes for structural element test in fire test

Design codes of structural element test for fire	Description
ASTM E119-11: Standard Test Methods for Fire Tests of Building Construction and Materials	Used in North America
ISO-834: Fire Resistance Test Elements of Building Construction	Used internationally
CAN/ULC-S101: Standard Methods of Fire Endurance Test of Building Construction and Materials	Used in North America
AS1530.4: Methods for Fire Tests on Building Materials, Components and Structures-Fire Resistance Test of Elements of Construction	Used in Australia
BS 476: Fire Tests on Building Materials and Structures	Used in European region
FRC Code: Code of Practice for Fire Resisting Construction 1996	Used in Hong Kong and was published by Building Departments of The Government of The Hong Kong Special Administrative Region
JIS A 1304: Method of Fire Resistance Test for Structural Parts of Buildings	Used in Japan
Uniform Building By-Laws 1984	Used in Malaysia

Different fire curve which is also known as time-temperature curve may exist. However, none of them can actually be used to represent the real fire because the real fire is more complex in behaviour and it depends on a number of variables such as the size of ventilation opening, quantity of combustible materials available and the lining materials present at the scene [30]. Hence, no two real fires are the same. Again, the development of fire curve is just a representative of the real fire and using scientific approach and under the expert jurisdiction; they are accepted as an indicator of the real fire.

Harmathy et al., investigated the severity of exposure in ASTM E119 (2010) and ISO-834 (2014) fire curve [31]. They tested the floor, wall and furnace by following the requirements in ASTM E119 (2010) and ISO-834 (2014). There were two major differences between these two curves. First, the thermocouples in the ASTM E119 (2010) test are protected in protective tubes while the thermocouples in ISO-834 (2014) furnace are bare. The second difference is that both standards have different time-temperature curves. ISO-834 (2014) curve can be described by the following formula:

$$\mathbf{T_f} = \mathbf{T_o} + 345\mathbf{log}_{10}(1 + 8t) \qquad (2.1)$$

where T_f is the average furnace temperature in °C, T_o is the initial temperature in °C and t is time in minutes. From their study they concluded that, if the structural elements were exposed for a short duration of up to 1.5 h, the ASTM E119 (2010) test give a more severe result than the ISO-834 (2014) test. But, after 1.5 h, the difference in severity of the specimen in the tests carried out using these two fire curves is negligible. The difference between the two curves can be seen in Fig. 2.3. Moreover, Eq. (2.1) is also used in BS 476 (1987). In Malaysia, design for fire safety follows Uniform Building By-Laws (1984) and the test for fire resistance is in accordance with BS 476 (1987). This will be discussed in subsequent subsection.

There is also a standard for hydrocarbon pool fire. The need to develop this type of temperature-time curve is due to the failure of the fire proof steel members that was exposed to petroleum spill in the late 1980s [32]. Therefore, in addition to ASTM E119 (2010), ASTM E1529: Standard Test Methods for Determining Effects of Large Hydrocarbon Pool Fires on Structural Members and Assemblies and UL 1709: Rapid Rise Fire Tests of Protection Materials for Structural Steel were established. These two fire curves deals with fire that spreads rapidly such as the hydrocarbon fire. The difference between ASTM E119 (2010) fire curve and ASTM E1529 (2016) fire curve are: (1) It will take just 5 min for the temperature to rise up to 1100 °C in ASTM E1529 (2016) but in ASTM E119 (2010) it will be only about 500 °C and (2) the installation of thermocouples in both tests. For ASTM E119 (2010), the thermocouples are required to be contained in protective capped steel whereas, for ASTM E1529 (2016), the thermocouples are bare [32]. The ASTM E119 (2010) curve is similar to the UL 263 (2011) whereas, the ASTM E1529 (2016) is similar to the UL 1709 (2011) as shown in Fig. 2.4.

Another hydrocarbon based fire curve is HCM for Modified Hydrocarbon Fire in French. In HCM, the temperature will be up to 1200 °C in just 10 min and after

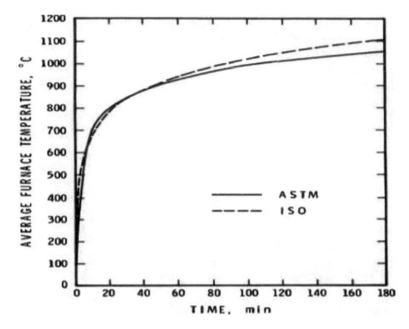

Fig. 2.3   Comparison between ASTM E119 (2010) and ISO-834 (2014) curves [31]

Fig. 2.4   Comparison between ASTM E119 (UL263) and ASTM E1529 (UL1709) standard fire curve [32]

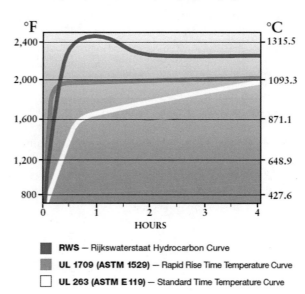

RWS – Rijkswaterstaat Hydrocarbon Curve

UL 1709 (ASTM 1529) – Rapid Rise Time Temperature Curve

UL 263 (ASTM E 119) – Standard Time Temperature Curve

that the rate will slow down. After 30 min the temperature will only rise to 1300 °C and remain constant until the end of the test [33]. Apart from above mentioned hydrocarbon based fire curve, there are also other hydrocarbon fire curves such as RWS (Netherlands), Hydrocarbon curve (Eurocode 1), RABT ZTV curve (German) and Dublin Port Tunnel (DPT) (EFNARC 2006).

2.2.3 Fire Test in Previous Research Studies

Previous research studies ran a fire test to structural elements under two different conditions, namely (1) stressed condition and (2) unstressed condition. Stressed and unstressed refers to specimens condition during heating process. In stressed condition, specimens are loaded simultaneously during heating process. While for unstressed condition, specimens were heated without loading. The specimens were loaded at room temperature after cooling down process. In this section, these two condition will be discussed using experimental work by Lu et al. for stress condition and for unstressed condition [10, 34].

Lu et al. ran fire test using stressed condition using gas furnace [10]. The test set up is shown in Fig. 2.5. The fire test set up consist of gas furnace, load reaction frame, loading system, fire temperature control unit and data acquisition system. Fire control unit controls the temperature of the furnace according to fire-temperature curves. The temperatures within the specimen were recorded by thermocouples that were mounted inside the specimens during casting process. The specimens were bolted to steel cylinder which was connected to reaction frame on both side. Loading was applied to the specimens using hydraulic jack. The specimens were loaded to designated load level before heating.

As for unstressed condition, Yaqub and Bailey [34], use electric furnace with dimension of 1.6 m × 1.2 m × 1.5 m as shown in Fig. 2.6. The temperature within the furnace were controlled by two type K thermocouples mounted at mid-height and at the top of furnace. The temperature of specimens were measured by attaching type K thermocouple on the surface of each columns. The specimens were left to cool down to room temperature before being subjected to axial compression test.

2.2.4 Fire Provision of Concrete-Filled Double Skin Steel Tubular (CFDST) Columns in Standards/Codes

Currently, the provision of fire safety for CFDST columns does not exist in Standards or Codes. The closest possible cross section that can be used in designing CFDST is CFST. Even before CFST was included in codes, designer used the provision of bare steel columns to cater for CFST which resulted in over design. A study by Han et al. proved that the design method of fire protection in Chinese Code (GB50045-95)

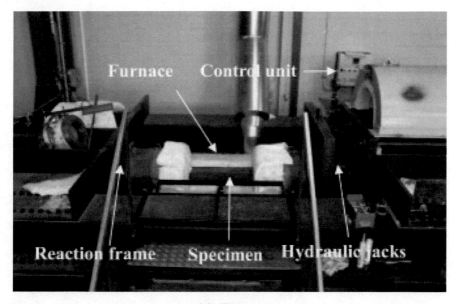

(a) Test set up

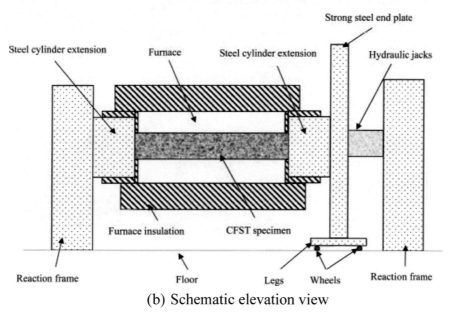

(b) Schematic elevation view

Fig. 2.5 Test set up **a** test set up, **b** schematic elevation view [10]

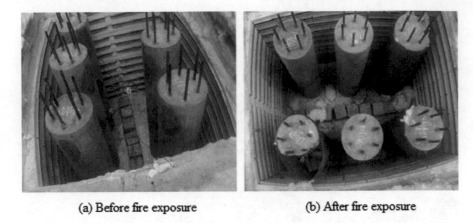

(a) Before fire exposure   (b) After fire exposure

Fig. 2.6 Electric furnace [34]

for bare steel columns cannot be used for CFST [21]. Subsequently, much research studies have been carried out only on the behavior of CFST when subjected to fire. As a result of such studies, design equation that can be used when designing CFST columns were developed and incoporated into codes (e.g., Eurocodes 4: Design of Composite Steel and Concrete Structures, EC4). However, EC4 only caters for three types of composite columns which are; (a) concrete encased profiles; (b) partially encased profiles and (c) concrete filled profiles as shown in Fig. 2.7. Encased profiles are usually preferred in Europe and United State of America (USA), whereas concrete filled profile or concrete-filled steel tubes are the preferable choice in the Far East [35].

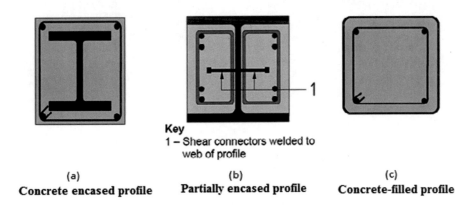

Key
1 – Shear connectors welded to
web of profile

(a)   (b)   (c)
Concrete encased profile   **Partially encased profile**   **Concrete-filled profile**

Fig. 2.7 Typical cross section in Eurocodes 4 **a** concrete encased, **b** Partially encased and **c** concrete-filled (European Committee for Standardization 2008)

2.2.5 Fire Provision in Malaysia

In Malaysia, the provision of fire safety is in accordance with Uniform Building By-Laws (1987). Every element in a building are given fire resistance period according to (1) type of structures i.e., columns, beams and walls, (2) types of building and (3) types of material used in the building i.e., reinforced concrete, steel and masonry.

In By-Laws (1987), type of building is classified into two categories, (1) building with single staircase and (2) building other than single storey building. The latter is than classified into another eight purpose group i.e., small residential, shop and office. For building with single staircase, it is clearly stated in By-Laws 194 (a) that each element of structure should at least achieved the fire resistance of one hour. Meanwhile, for building other than single storey building, the minimum period of fire resistance is based height, floor area and cubic capacity stated in Ninth Schedule of By-Laws (1987).

By-Laws also stated fire resistance period according to the types of element. Different types of material used is required to have different fire resistance. For instance, the fire resistance period for reinforced concrete columns ranging from half an hour to four hours depended on construction and materials as well as dimension of the concrete column stated in Ninth Schedule of By-Laws (1987). Larger column result in longer fire resistance period.

There is no provision for CFDST columns in By-Laws (1987). However, By-Laws (1987) do have provision for encased steel stanchions in Ninth Schedule. Similar to reinforced concrete columns, the fire protection period is based on construction and material and minimum thickness of protection. Protection is classified further into a number of construction material such as concrete, solid bricks, solid blocks, sprayed asbestos and gypsum plasterboard. Based on minimum thickness of protection, the fire resistance period can be as short as half an hour and as long as 4 h.

2.3 Behavior of Concrete-Filled Double Skin Steel Tubular Columns Exposed to Fire

Fire resistance of hollow structural steel (HSS) in high rise building without fire protection is usually less than half an hour [36]. According to fire safety design, the required fire resistance rating depends on type of building which could range from 30 min to 3 h [37]. Filling HSS with concrete is considered as one of the method to increase fire resistance of HSS [19]. Moreover, concrete is known to possess low thermal conductivity and high heat capacity thus acting as heat sink. In the case of CFDST columns, concrete acts as additional fire protection to the inner tube.

Figure 2.8 shows schematic view of typical behavior of CFST and CFDST columns during fire. The columns are loaded during the whole course of fire. During early stage of fire, outer steel tube carries most of the load. Due to fire, outer steel tube expands faster than concrete and inner steel tube. At the same time, mechanical

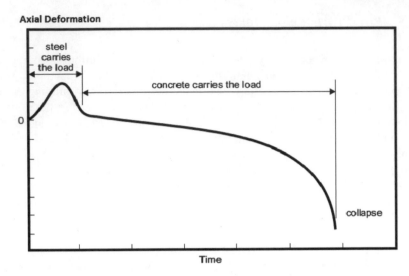

Fig. 2.8 Typical behavior of CFST/CFDST columns exposed to fire [37]

properties of outer steel tube starts to degrade and this leads to strength deterioration of outer steel tube. After reaching its yield stress, the outer steel tube started to lose its bearing capacity. Load is transferred to concrete in the case of CFST and both concrete and inner steel tube for CFDST. At this moment, the temperature of inner steel tube is quite low due to protection offered by surrounding concrete. This enables the columns to carry loads for a period of time before the columns finally collapse due to deterioration of strength of the whole structure. The fire resistance of a column is defined as the total time from the start of fire until the collapse of the column [37].

However, in some cases, CFDST columns are heated without loading. This condition is considered to be more conservative especially when residual strength of the columns is the main concern [38]. The importance of knowing residual strength is; (1) to assess the potential damage caused by fire and (2) to calculate the minimum post fire repair needed [39, 40]. Since CFDST columns consist of two different materials, i.e., concrete and steel, both material characteristic when exposed to fire are discussed individually in subsequent sections. After that, the post-fire behaviors of CFST are discussed.

References

1. Shakir Khalil, H.: Composite columns of double-skinned shells. J. Constr. Steel Res. **19**(2), 133–152 (1991)
2. Tao, Z., Han, L.H., Zhao, X.L.: Behaviour of concrete-filled double skin (CHS inner and CHS outer) steel tubular stub columns and beam-columns. J. Constr. Steel Res. **60**(8), 1129–1158

(2004)

3. Zhao, X.L., Han, B., Grzebieta, R.H.: Plastic mechanism analysis of concrete-filled double-skin (SHS inner and SHS outer) stub columns. Thin-Walled Struct. **40**(10), 815–833 (2002)

4. Zhao, X.L., Grzebieta, R.: Strength and ductility of concrete filled double skin (SHS inner and SHS outer) tubes. Thin-Walled Struct. **40**(2), 199–213 (2002)

5. Uenaka, K., Kitoh, H., Sonoda, K.: Concrete filled double skin circular stub columns under compression. Thin-Walled Struct. **48**(1), 19–24 (2010)

6. Han, L.H., Li, Y.J., Liao, F.Y.: Concrete-filled double skin steel tubular (CFDST) columns subjected to long-term sustained loading. Thin-Walled Struct. **49**(12), 1534–1543 (2011)

7. Muhammad, N.B., Fan, J.S., Nie, J.G.: Effects of hollowness on strength of double skinned concrete filled steel tubular columns of different geometries under axial loading. Appl. Mech. Mater. **94–96**, 1746–1751 (2011)

8. Hassanein, M.F., Kharoob, O.F.: Compressive strength of circular concrete-filled double skin tubular short columns. Thin-Walled Struct. **77**, 165–173 (2014)

9. Lu, H., Han, L.H., Zhao, X.L.: Fire performance of self-consolidating concrete filled double skin steel tubular columns: experiments. Fire Saf. J. **45**(2), 106–115 (2010)

10. Lu, H., Zhao, X.L., Han, L.H.: Testing of self-consolidating concrete-filled double skin tubular stub columns exposed to fire. J. Constr. Steel Res. **66**(8–9), 1069–1080 (2010)

11. Yang, H., Han, L.H., Wang, Y.C.: Effects of heating and loading histories on post-fire cooling behaviour of concrete-filled steel tubular columns. J. Constr. Steel Res. **64**(5), 556–570 (2008)

12. Lu, H., Zhao, X.L., Han, L.H.: FE modelling and fire resistance design of concrete filled double skin tubular columns. J. Constr. Steel Res. **67**(11), 1733–1748 (2011)

13. Imani, R., Bruneau, M., Mosqueda, G.: Simplified analytical solution for axial load capacity of concrete-filled double-skin tube (CFDST) columns subjected to fire. Eng. Struct. **102**, 156–175 (2015)

14. Campione, G., et al.: Strength of hollow circular steel sections filled with fibre-reinforced concrete. Can. J. Civ. Eng. **27**(2), 364–372 (2000)

15. Han, L.., Huo, J.S., Wang, Y.C.: Compressive and flexural behaviour of concrete filled steel tubes after exposure to standard fire. J. Constr. Steel Res. **61**(7), 882–901 (2005)

16. Tao, Z., Han, L.H.: Behaviour of fire-exposed concrete-filled steel tubular beam columns repaired with CFRP wraps. Thin-Walled Struct. **45**(1), 63–76 (2007)

17. Lie, T.T., Chabot, M.: A method to predict the fire resistance of circular concrete filled hollow steel columns. J. Fire. Prot. Eng. **2**(4), 111–124 (1990)

18. Lam, D., Williams, C.A.: Experimental study on concrete filled square hollow sections. Steel Compos. Struct. **4**(2), 95–112 (2004)

19. Kodur, V.K.R.: Guidelines for fire resistance design of concrete-filled steel HSS columns-state-of-the-art and research needs. Steel Struct. **7**, 1–10 (2007)

20. Kodur, V.K.R., Lie, T.T.: Experimental Studies on the Fire Resistance of Circular Hollow Steel Columns Filled with Steel-Fibre-Reinforced Concrete, p. 1995. National Research Council Canada, Institute for Research and Construction (1995)

21. Han, L.H., Zhao, X.L., et al.: Experimental study and calculation of fire resistance of concrete-filled hollow steel columns. J. Struct. Eng. **129**(3), 346–356 (2003)

22. Elchalakani, M., Zhao, X.-L., Grzebieta, R.: Tests on concrete filled double-skin (CHS outer and SHS inner) composite short columns under axial compression. Thin-Walled Struct. **40**(5), 415–441 (2002)

23. Ding, J., Wang, Y.C.: Realistic modelling of thermal and structural behaviour of unprotected concrete filled tubular columns in fire. J. Constr. Steel Res. **64**(10), 1086–1102 (2008)

24. Baldwin, R.: The analysis of fire safety. Accid. Anal. Prev. **6**(3–4), 205–222 (1974)

25. Chowdhury, E.U.: Behaviour of Fibre Reinforced Polymer Confined, pp. 1–235. Department of Civil Engineering. Canada, Queen's University. Doctor of Philosophy (2009)

26. Purkiss, J.: Fire Safety Engineering Design of Structures. Butterworth-Heinemann, Jordan Hill, Oxford (1996)

27. Liu, L.: Fire Performance of High Strength Concrete Materials and Structural Concrete, pp. 1–229. Faculty of The College of Engineering and Computer Science. Florida, Florida Atlantic University. Doctor of Philosophy (2009)

28. Shields, T., Silcock, G.W.: Buildings and Fire, pp. 1–452. Longman Scientific & Technical, New York (1987)
29. Wang, Y.C.: Steel and Composite Structures. Spon Press, London (2002)
30. Wang, Y.C.: Some considerations in the design of unprotected concrete-filled steel tubular columns under fire conditions. J. Constr. Steel Res. **44**(3), 203–223 (1997)
31. Harmathy, T.Z., Sultan, M.A., MacLaurin, J.: Comparison of severity of exposure in ASTM E 119 and ISO 834 fire resistance tests. J. Test. Eval. **15**(6), 371–375 (1987)
32. Milke, J., Kodur, V., Marrion, C.: Overview of Fire Protection in Buildings (2002). In Online: http://www.fema.gov
33. Pimienta, P., et al.: Fire protection of concrete structures exposed to fast fires. In: Lonnermark, A., Ingason, H. (eds.) Fourth International Symposium on Tunnel Safety and Security, pp. 235–247. SP Technical Research Institute of Sweeden, Frankfurt am Main, Germany (2010)
34. Yaqub, M., Bailey, C.G.: Repair of fire damaged circular reinforced concrete columns with FRP composites. Constr. Build. Mater. **25**(1), 359–370 (2011)
35. Collings, D.: Steel-Concrete Composite Buildings: Designing with Eurocodes. Thomas Telford, London (2010)
36. Han, L.H., Li, W., Bjorhovde, R.: Developments and advanced applications of concrete-filled steel tubular (CFST) structures: members. J. Constr. Steel Res. **100**, 211–228 (2014)
37. Zhao, X.L., Han, L.H., Lu, H.: Concrete-Filled Tubular Members and Connections. Spon Press, London and New York (2010)
38. Liu, F., Gardner, L., Yang, H.: Post-fire behaviour of reinforced concrete stub columns confined by circular steel tubes. J. Constr. Steel Res. **102**, 82–103 (2014)
39. Hertz, K.D.: Concrete strength for fire safety design. Mag. Concr. Res. **57**(8), 445–453 (2005)
40. Han, L.H., Yang, H., Cheng, S.L.: Residual strength of concrete filled RHS stub columns after exposure to high temperatures. Adv. Struct. Eng. **5**(2), 123–134 (2002)

Chapter 3
Post-fire Behavior of Concrete-Filled Double Skin Steel Tubular Columns

In order to understand the behavior of post-fire behavior of composite column, i.e., CFST and CFDST columns, the behavior of its different component needed to be properly studied first.

3.1 Concrete

The mechanical properties of concrete, i.e., compressive strength, tensile strength and modulus of elasticity of the concrete decreased with the increasing temperature [1–3]. Factors that significantly influence compressive strength at high temperature are compressive strength at ambient temperature, rate of heating and presence of binder in concrete mix such as silica fume, fly ash and slag [3]. Tensile strength of concrete in normal strength concrete (NSC) is about 10% of its compressive strength. However, it is an important property since crack occurs in concrete when it is under tensile stress. Tensile strength of the concrete becomes more crucial at high temperature as occurrence of fire induced spalling. Factors influencing tensile strength of the concrete at high temperature is similar to compressive strength mentioned before [3]. Factors affecting modulus of elasticity at high temperature are moisture loss, high temperature creeps and type of aggregates. At high temperature, hydrated cement paste disintegrated and the bond in concrete microstructures disintegrates leading to reduction in concrete modulus of elasticity [3]. In addition to that, the presence of stress during high temperature exposure also can significantly affect mechanical properties of concrete [1, 2]. Liu [2] presented a table which is convenient for classifying the effect of temperature on mechanical properties of concrete Table 3.1.

The mechanical properties test of the concrete can be carried out in two ways, namely, (1) high temperature test and (2) residual test. High temperature test was done by way of loading and exposing the specimens to fire simultaneously while in residual tests, the specimens were loaded after the specimens were allowed to cool down to

© The Author(s), under exclusive license to Springer Nature Singapore Pte Ltd. 2021 19
K. K. Choong et al., *Concrete-Filled Double Skin Steel Tubular Column with Hybrid Fibre Reinforced Polymer*, SpringerBriefs in Applied Sciences and Technology, https://doi.org/10.1007/978-981-16-2715-6_3

Table 3.1 Condition of Portland cement concrete during heating process [2]

Temperature (°C)	Behavior of concrete
250–420	Some spalling may take place, with pieces of concrete breaking away from the surface
300	Strength loss starts, but in reality only the first few centimeters of concrete exposed to a fire will get any hotter than this, and internally the temperature is still below 300 °C
550–600	Cement-based materials experience considerable creep and lose their load bearing capacity
600	Above this temperature, concrete no longer function at its full structure capacity
900	Air temperature in fires rarely exceeds this level, but flame temperature can rise to 1200 °C and beyond

room temperature after exposure to fire [1]. It was observed that when heating an element while loading, the loss in mechanical properties is less than unloaded element [2, 4]. Therefore, exposing the specimens in an unstressed condition result in more conservative value which leads to safer design [5]. Hertz [6] suggested that it is safer to use the value from unstressed test specimens since stressed concrete subjected to high temperature is 25% stronger than unstressed concrete.

When concrete is subjected to high temperature, cracking is one of the main concerns; starting from minor hairline crack that are sometimes not visible over the concrete surface to major crack such as fire induced spalling. Due to porous nature of concrete, it allows water to permeate slowly into the concrete. Once heated, water turns to water vapor which expands faster than it can escape from the concrete. This leads to built up of pressure inside the concrete. When the stress induced by such pressure is greater than concrete tensile strength, concrete starts to cracks internally. These microcracks may not be visible from the outside of the concrete. Furthermore, concrete may not break for an extended period since formation of cracks releases the trapped water vapor thus the pressure as well. Overtime, more pressure is built up inside the concrete which eventually causes the concrete to crack [7].

Due to weakening of concrete after prolong exposure to high temperature, a chunk of concrete slough off from the surface of the concrete. This so called spalling sometimes occurred during early stage of fire and sometimes accompanied by violent explosion [3, 6, 8]. The latter was caused by rapid heating of concrete that creates thermal gradient between surface and core concrete. Spalling is known to occur in all types of concrete. However, it is more frequent in high strength concrete (HSC) due to its dense microstructures [3]. Spalling needs to be avoided in case of fire because it exposed deeper layer of concrete or/and reinforcement thus, allowing heat to transfer further into the structure. Once this happen, the structures strength will be greatly reduced [3, 8].

Table 3.2 shows the effect of temperatures on composition of concrete. This condition was observed after the concrete cools down to room temperature. This Table specifically relate to changes in composition of concrete under different temperatures.

Table 3.2 Effect of temperatures on composition of concrete [6, 10]

Temperature (°C)	Changes in composition of concrete
150	Chemically bound water is released from the hydrated calcium silicate
300	Formation of microcracks through out concrete due to dehydration of cement matrix (C-S-H gel)
400–600	Decomposition of calcium hydroxide into calcium oxide and water
Above 600	Decomposition of hydrated calcium silicate
Above 1150	Feldspar melts and the remaining minerals of cement paste turn into glass phase

Hertz [6] and Chang et al. [9] suggest that concrete that was exposed to temperature under 300 °C still retain 90% of its unheated compressive strength. Furthermore, during cooling down process, concrete absorbed moisture from the surrounding air leading to strength recovery after fire. Chan et al. [10] classified three temperature ranges affecting concrete. Firstly, temperature range which is below 400 °C, where only small part of unheated compressive strength is lost. As studied by Chan et.al [10] and Chang et al. [9], when concrete was exposed to temperature below 400 °C, the lost in compressive strength is only 10% and 15%, respectively. Generally, 300 °C is set as the threshold value, above which significant strength lost of concrete occurs. Concrete that was exposed to temperature around 300 °C changes color to reddish tint (pink to red) due to oxidation of iron in aggregates [4, 8]. Second temperature range that was classified by Chan et al. [10] is 400–800 °C. The lost of strength during this range is regarded as critical where the strength is reduced to only 15% of the ambient temperature compressive strength at temperature 800 °C [9]. If the concrete was exposed to temperature of more than 800 °C, the concrete structure is considered as structurally damaged since only about 9–15% of its ambient temperature compressive strength is left [10]. Moreover, during cooling process, calcium oxide absorbs water from air and expands. This process causes the cracks to open up even more thus reducing residual strength of concrete [6]. Therefore, in accessing the residual strength of fire-damaged concrete, Hertz [6] advised that it be done at least a week after fire exposure, depending on the geometry of the structures.

However, the reduction of concrete strength depends not only on exposure temperature. It also depends on exposure time. Longer exposure time leads to larger reduction in residual strength. Nevertheless, this hypothesis is only true for the first two hours of exposure. The study conducted by Mohamedbhai [11] indicates that exposure time significantly affects residual strength of the concrete. However, this effect diminished as exposure temperature and exposure time increased. This later relation was confirmed by Yang et al. [12].

As mentioned above, color changes in concrete can be taken as an indicator of temperature history of fire-exposed concrete. This is because, concrete that changed in color during fire exposure retained it color after being cooled off. Changes in color are due to presence of iron in aggregates. As shown in Table 3.3 the color changes as the temperature increased. Therefore, by using visual observation, engineers can

Table 3.3 Color changes in concrete at high temperature [4, 8]

Temperature (°C)	Concrete color
300–500	From normal to pink (or reddish)
500–900	Whitish grey
900–1000	Buff

roughly predict the temperature history of the fire-damaged concrete. With that, appropriate repair measure can be proposed in advance. For concrete that was exposed to around 300 °C (pink to red discoloration), the affected concrete is knocked down and replaced with the new one. The load bearing capacity of the structure can be improved by increasing the cross section with a new layer of concrete or additional reinforcement, e.g., steel, fiber reinforce polymer (FRP). This is the common repair method for beams and columns [13]. Repair of fire-damaged columns is discussed in detail in subsequent section.

3.2   Steel

Both yield strength and modulus of elasticity of the steel show great variation when exposed to fire. This happened due to variation of composition in steel [1] Generally, yield strength and modulus of elasticity decreased with increasing temperature. However, if the maximum temperature of the steel is in between 500 and 600 °C, yield strength of the steel remained almost unchanged [1, 14]. In contrast, Eurocodes 3 stated that yield strength of carbon steel decreased to about half of its room temperature strength at approximately 600 °C (Fig. 3.1). Gunalan and Mahendran [15] studied the yield strength of low and high grade steel subjected to different temperature. The temperatures are in the range of room temperature to 800 °C. The results showed that high grade steel loses its strength more rapidly than low grade steel in the range of 500–600 °C exposure. In addition to that, in a study conducted by McCann et al. [16], there seem to be very small reduction in ultimate strength when cold-formed steel materials were exposed up to 400 °C. These variations in temperatures are due to difference in composition in steel. The reduction in ultimate strength was very small and can be neglected.

As for modulus of elasticity, no significant effect can be observed when the steel were heated to 250 °C. However, the modulus of elasticity decreased gradually when steel was exposed to temperature above 250 °C [17] Qiang et al. [14] observed that when high strength steel (steel grade S460 and S690) were exposed to temperatures of 800 and 1000 °C, it started to gradually lose its modulus of elasticity. However, both steel grades regain more than 75% and 60% of its modulus of elasticity, respectively, after cooling process. Similar result was found in a study by Gunalan and Mahendran [15] for high grade steel (S690) where, modulus of elasticity starts to decrease gradually at temperature of 800 °C. The modulus of elasticity of the high grade steel remains unchanged up to temperature of approximately 400 °C. In a

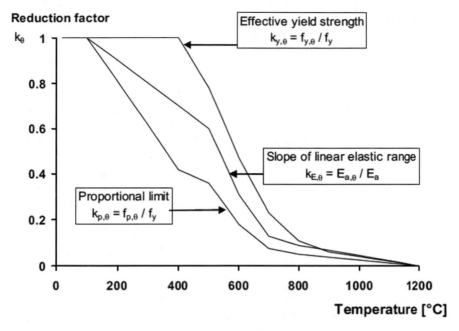

Fig. 3.1 Reduction factors for the stress-strain relationship of carbon Steel at elevated temperatures (Eurocodes 3)

recent research study by Lu et al. [18], it was observed that modulus of elasticity of high grade steel (Q420, hot-rolled steel) was unaffected when it was exposed to temperature below 700 °C. As for low grade steel (S460), modulus of elasticity remained unaffected up until 700 °C with only 4% reduction when the temperature was 800 °C [15]. This result was in close agreement with that reported by Lu et al. [18], where low grade steel (Q235 and Q345) regain 100% of its original modulus of elasticity after exposure to temperature of 800 °C. This indicates that low grade steel still retains its stiffness after exposure to high temperature.

The critical temperature of the steel is stated as 550 °C. If the temperature goes beyond this critical temperature, the steel structures are considered unsafe [19]. In other words, if the exposure temperature is less than 550 °C, the yield strength of steel structures will be recovered upon cooling. However, if the exposure temperatures is in excess of this critical temperature of 550 °C, the yield strength will not be recovered after the cooling process [20]. In addition, according to ASTM E-119 (2010), critical temperature of steel is 538 °C (ASTM 2010).

3.3 Concrete-Filled Steel Tubular (CFST) Columns and Concrete-Filled Double Skin Steel Tubular (CFDST) Columns

So far, CFDST column was used in outdoor construction even though it was more superior to CFST. This might be due to lack of fire related study of CFDST that leads to uncertainty in using this type of column. Therefore, in depth study particularly post-fire behavior of CFDST columns is needed. Knowledge about post-fire behavior or residual strength of fire-damaged CFDST columns is important; (1) to assess the extent of damage caused by fire, (2) to establish approach to determine the structural fire protection for minimum post-fire repair [21] and (3) to determine the amount of repair needed for post-fire repair. Table 3.4 summarizes research studies done on CFDST columns exposed to fire. However, research studies that deals with post-fire behavior of CFDST columns is very limited.

3.3.1 Residual Strength, Stiffness and Ductility of Fire-Damaged CFST Columns

After exposure to ISO-834 standard fire curve for 180 min, circular and square columns tested in a study by Tao et al. [28] suffered significant loss in strength and stiffness as shown in Fig. 3.2 and Fig. 3.3, respectively (CSC represent unheated circular CFST and CSCF-0 represent heated circular CFST column). Circular CFST columns lost 52.3% of its ambient temperature strength after 180 min of exposure;

Table 3.4 Summary of research conducted on concrete-filled double skin steel tubular columns exposed to fire

Researchers	Focus of research
Imani et al. [22]	Developed analytical procedure to calculate axial load capacity of CFDST column subjected to fire
Lu et al. [23]	Developed numerical model using finite element to analyze the fire behavior of CFDST columns exposed to fire
Lu et al. [24]	Fire performance of CFDST columns filled with self-consolidating concrete (SCC) exposed to ISO-834 standard fire
Lu et al. [25]	Behavior of CFDST stub columns filled with SCC and fiber reinforced SCC (steel and polypropylene fiber) exposed to AS 1530.4 standard fire
Yang and Han [26]	Developed fire resistance and fire protection material thickness formulae of CFDST columns. Determination of temperature distribution within column section was made using finite element method. The columns were exposed to ISO-834 standard fire
Yang and Han [27]	Developed theoretical model to determine deformation and strength of CFDST beam-columns exposed to ISO-834 standard fire

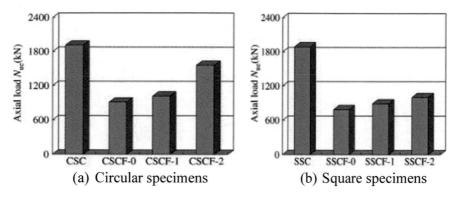

(a) Circular specimens   (b) Square specimens

Fig. 3.2   Ultimate strength of concrete-filled steel tubular columns [28]

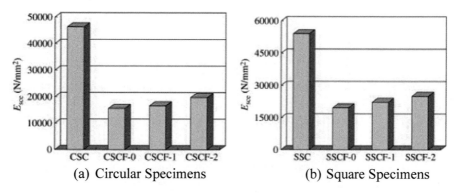

(a) Circular Specimens   (b) Square Specimens

Fig. 3.3   Stiffness of concrete-filled steel tubular columns [28]

whereas square CFST columns lost additional 6.1%. Furthermore, CFST suffered greater loss in stiffness when compared with strength where for circular CFST columns, the corresponding loss is as high as 66.3% and for square CFST columns, the loss is also high at 63.7%. From Fig. 3.4, it can be seen that post-peak curve of unheated CFST column (CSC and SSC) is steeper than fire-damaged CFST column (CSCF-0 and SSCF-0). This proved that once heated, CFST columns lost its stiffness as mentioned above. Moreover, strain corresponding to ultimate load is greatly increased for fire-damaged CFST columns. This mean that fire-damaged CFST column is more ductile than original unheated column. Both circular and square CFST columns possess the same characteristics.

Similar observation was reported by Tao and Han [29]. Tao and Han [29] carried out a study on fire exposed CFST beam-columns both for circular and square cross-section. CFST beam-columns were exposed to ISO-834 standard fire curve for 180 min with varying parameters such as slenderness and load eccentricity. In order to quantify the effect on strength of fire exposed CFST beam-columns, Tao and Han [29] used Residual Strength Index (RSI) suggested by Han and Huo [30], where

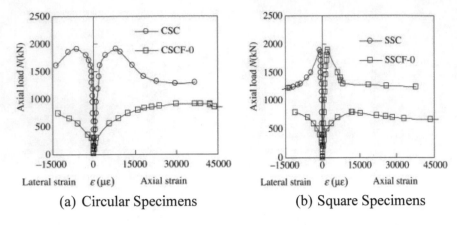

(a) Circular Specimens   (b) Square Specimens

Fig. 3.4 Load-strain curves for fire-exposed concrete-filled steel tubular columns [28]

$$RSI = \frac{N_u - N_u(t)}{N_u} \qquad (3.1)$$

$N_u(t)$ is residual strength of fire exposed CFST columns and N_u is ultimate strength of CFST columns at ambient temperature. RSI of beam-columns studied by Tao and Han [29] are 69.7% and 65% for circular CFST beam-columns and square CFST beam-columns, respectively.

In a recent research study by Rush et al. [31] CFST columns were subjected to 120 min (unprotected columns) and 180 min (protected columns) of exposure time. Protected and unprotected column refers to the columns with and without fire protection layers, respectively. It was observed that unprotected CFST columns were able to retain between 40 and 60% of its original strength. The observation is similar to that reported Tao et al. [28] and Tao and Han [29].

Stiffness of CFST columns can be measured by calculating secant stiffness, k from load-displacement curve as shown in Fig. 3.5. Similar method has been adopted by Tao and Han [29]. It is apparent that fire exposure result in large reduction in stiffness of CFST beam-columns.

Apart from residual strength and stiffness of CFST columns, ductility is also one of main parameter needed to be investigated after fire exposure. Liu et al. [5] used Ductility Index (DI) in order to calculate ductility of fire-damaged CFDST.

$$DI = \frac{\Delta_{0.85}}{\Delta_u} \qquad (3.2)$$

$\Delta_{0.85}$ and Δ_u are defined in Fig. 3.6.

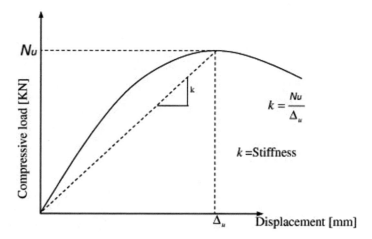

Fig. 3.5 Secant stiffness [32]

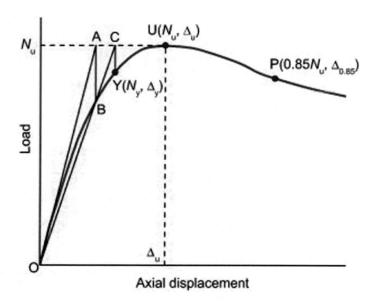

Fig. 3.6 Parameters in determining ductility index of concrete-filled steel tubular and concrete-filled double skin steel tubular columns [5]

3.3.2 Cross-Sectional Dimension

Yin et al. [33] conducted a parametric study on circular and square CFST columns exposed to high temperature and the result were compared between circular and square columns with similar cross section area and steel tube thickness. The comparison showed that, circular columns performed better than square when subjected to high temperature. This is due to the overall perimeter of square columns which is relatively longer. When all side of square columns was exposed to fire, they absorb more heat than circular counterparts due to its longer perimeter thus resulting in higher temperature within the columns cross-section. In a previous study, Kodur and Mackinnon [34] also found out that circular CFST performed better than square CFST. The corner of the square CFST column was found to be heated faster than its flat surfaces. Therefore, concrete at the corner degrades faster. CFST columns with circular sectional shape were found to be able to withstand the longest exposure time before failure than elliptical, square and rectangular sectional shape. All shape posses similar strength and steel section area [35].

This is quite contradictory with statement by Han et al. [36] where the authors found that cross-sectional profile was one of the main significant influence for distribution of temperature in the columns. For instance, larger sectional dimension means columns possess greater mass of concrete that eventually slows down the temperature rise in the columns. In addition, larger columns lead to larger area of concrete with lower temperature. Hence, the concrete can carry loads throughout the fire course [37, 38].

In a parametric study conducted by Yang and Han [27], outer steel tube diameter was found to significantly affect the fire resistance of CFDST columns as shown in Fig. 3.7 where D_o is the diameter of outer steel, t_R is fire resistance and λ is slenderness ratio as defined in Yang and Han [27]. The increased in fire resistance with the increment of outside diameter can be attributed to longer time for the temperature to reach concrete and inner tube to their limiting temperature. Similar results were found in the parametric study conducted by Lu et al. [24]. Moreover, longer fire resistance eventually lead to higher post-fire strength of the columns. Column that was designed with longer fire resistance was able to withstand longer fire exposure. Therefore, with similar fire exposure time it can result in higher residual strength.

Lu et al. [24] and [25] conducted study on CFDST columns exposed to AS1530.4 and ISO-834 standard fire curves, respectively. Both studies found that outer steel tube perimeter had significant influence on the temperature distribution of CFDST columns. In Lu et al. [24], two CFDST columns with different parameter of outer

Fig. 3.7 Effects of outer diameter on fire resistance of concrete-filled double skin steel tubular columns [27]

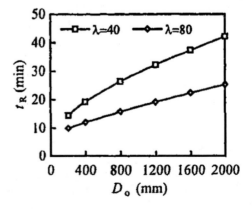

tube, SC1 and CC2, were compared. SC1 with outer tube perimeter of 1120 mm over CC2 with only 924 mm outer tube perimeter possess longer fire resistance. It is found that longer fire resistance lead to higher residual strength of CFDST columns. Lu et al. [25] extracted 10 and 15 min of temperature distribution during the test and plotted the temperature against outer tube perimeter of CFDST columns as shown in Fig. 3.8. T1, T2 and T3 are thermocouple location, where T1 measured outer steel tube temperature, T2 measured temperature of concrete and T3 measured inner steel tube temperature. From Fig. 3.8, it is clear that temperature is lower with larger perimeter.

3.3.3   Exposure Time

Han and Huo [30] exposed 12 CFST to fire according to ISO-834 standard in order to analyze the effect of different parameters on residual strength. One of the tested parameter was exposure time. CFST columns were exposed to fire for 90 min and the RSI was calculated. After that, parametric study was carried out to investigate a wider range of exposure time. The study found that increased fire exposure time leads to increase deflection and reduction in strength of CFST columns as can be seen in Fig. 3.9. In addition, increased exposure time also significantly affected RSI. However, when CFST columns were exposed for less than 10 min of fire, no reduction in strength can be detected. Tao and Han [29] exposed circular and square CFST beam-columns to fire according to ISO-834 standard. After 180 min exposure without fire protection, the circular beam-column residual strength index (RSI) of CFST, as defined by Han and Huo [30], is 69.7% and RSI for square CFST beam-column is 65%.

3.3.4   Temperature Distribution of Concrete-Filled Double Skin Steel Tubular (CFDST) Columns

As mentioned in previous section, the maximum temperature attained by steel and concrete is important in determining the extent of damage due to fire. In addition, changes in appearance, i.e., color, deformation can be taken as an indicator for temperature history of CFDST exposed to high temperature. Critical temperature is often used when defining failure of steel structural members. Critical temperature is defined as the temperature where steel begin to lose 50% of its room temperature strength [39]. Critical temperature for steel column is 538 °C (ASTM E119, 2010). As agreed by many researchers, critical temperature of concrete is in between 300 and 400 °C.

Lu et al. [24] and [25] exposed CFDST columns to fire according to ISO-834 and AS 1530.4 standard, respectively. All columns were tested to failure and were loaded and heated simultaneously. In order to measure temperature of each element

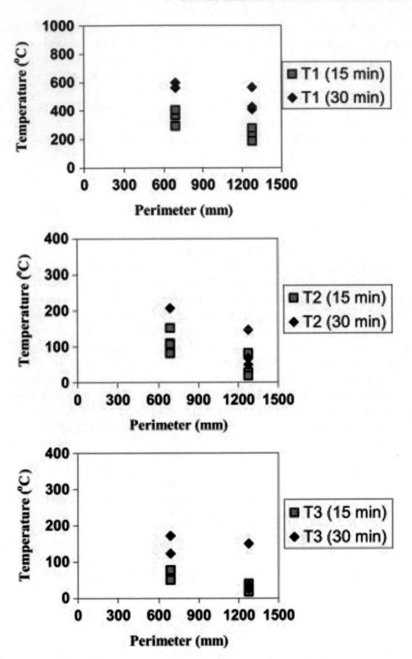

Fig. 3.8 Effects of outer tube perimeter on temperatures in concrete-filled double skin steel tubular column [24]

Fig. 3.9 Axial load versus mid-span lateral deflection curves for different fire exposure time [30]

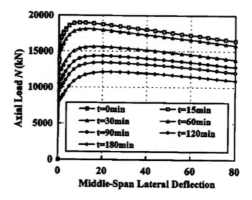

of CFDST columns, the researchers inserted thermocouple at different location corresponding to the different element of CFDST columns, i.e., outer steel tube, concrete and inner steel tube. From thermocouple readings, it was found that the temperature distribution within CFDST columns were not uniform. During early stage of exposure, temperature of outer steel tube rises rapidly. At around 100 °C, there was a relatively stable stage in concrete temperature. This phenomenon is observed by both researchers. This can be attributed to the heat being used to change the state of water to vapor instead of raising the temperature of the concrete. At the end of the test, the highest temperature gained by inner steel tube is between 59 and 197 °C when the outer steel tube temperature varies from 400 to 963 °C [25]. On the other hand, in the study by Lu et al. [24] the maximum temperature attained by outer steel tube and inner steel tube was found to be 940 °C and 450 °C, respectively. The difference in maximum temperature is due to the difference in cross sectional dimension, thickness of concrete and duration of fire exposure time.

From the above studies it can be seen that, the temperature distribution of inner steel tube did not reach critical temperature of steel, which is 538 °C. Therefore, inner steel tube together with concrete is still able to carry its entire ambient temperature load even after exposure to fire. This leads to minimum post-fire repair of CFDST columns.

3.4   Repair of Fire-Damaged Concrete-Filled Steel Tubular (CFST) Columns Using Fiber Reinforced Polymer (FRP)

To date, there is very limited research study can be found on repair of fire-damaged CFDST columns using FRP can be found in literature. Therefore, literature review on repair of fire-damaged is focused on CFST columns since the failure mode of outer steel tube of CFST is similar to the failure mode of outer steel tube of CFDST

Table 3.5 Summary of research conducted on fire-damaged concrete-filled steel tubular columns repaired with fiber reinforced polymer

Researchers	Focus of research
Tao and Han [29]	Investigated fire-damaged circular and square CFST beam-columns repaired with CFRP. Strength, stiffness and ductility of CFST beam-columns were enhanced. However, the enhancement decreased with increased eccentricity and/or slenderness ratio. Authors suggested that for slender members or members subjected to large bending moments, other repair measures should be taken
Tao et al. [28]	Investigated fire-damaged circular and square stub CFST columns and CFST beams repaired with CFRP. It was found that load carrying capacity and longitudinal stiffness of repaired CFST increased with increasing layer of CFRP; whereas, ductility decreased with increased CFRP layers
Tao et al. [40]	Investigated seismic behavior of fire-damaged CFST beam-columns repaired with Carbon FRP (CFRP). Ultimate strength, stiffness and ductility were increased and the increment is proportionate to number of CFRP layers. However, strength, stiffness and ductility were not fully restored up to original ambient temperature strength

columns. Previous research studies on repairing of fire-damaged CFST columns using FRP is shown in Table 3.5.

Repairing or retrofitting of columns or structures can be achieved through several methods such as steel or FRP jacketing or wrapping and/or section enlargement [41]. Both methods are widely used in the retrofitting of reinforced concrete structures. However, researchers such as Tao et al. [29] and Tao and Han [29] have started to use Carbon FRP (CFRP) to retrofit CFST after exposure to fire. CFRP exhibits superior performance over other types of FRP due to its high tensile strength to weight ratios as well as high tensile modulus to weight ratios [42]. In addition to that, CFRP also possesses high fatigue strength and a very low coefficient of thermal expansion and in some cases, negative thermal expansion. With this characteristic, there is virtually no expansion of CFRP up to 300 °C. CFRP is also chemically inert and not vulnerable to corrosion or oxidation at temperature below 400 °C [42]. The most common FRP used is Glass FRP (GFRP). It is commonly used due to its low cost, high tensile strength and resistance to chemical and temperature [42]. GFRP also possess high ultimate tensile strain which is a favorable characteristic for ductility enhancement of fire-damaged CFDST columns [43]. Besides, using GFRP in repairing CFDST columns can avoid long term effect of galvanic corrosion that occurs when CFRP is in contact with steel [43].

Tao et al. [28] tested eight CFST stub columns, four specimens per circular and square cross-section. Specimens were classified into three categories, which are (1) undamaged specimens, i.e., control specimens, (2) fire-damaged and unrepaired specimens, and (3) fire-damaged and repaired specimens. Specimens from category (2) and (3) were exposed to fire in accordance to ISO-834 standard for 180 min. After that, the specimens from category (3) were repaired with 1 layer of Carbon Fiber Reinforced Polymer (CFRP) or two layers of FRP with CFRP as first layer and Glass

FRP (GFRP) as second layer. Repair of fire-damaged columns was achieved through externally wrapping the unidirectional CFRP and GFRP sheet in hoop direction using hand lay-up method. All specimens were subjected to axial compression tests. The strength and stiffness enhancement due to CFRP wrapping is shown in Figs. 3.2 and 3.3. It seems that the enhancement both in strength and stiffness increased with increasing layers of CFRP. However, the increment in strength is more obvious than stiffness. In addition to that, the enhancement in strength is more significant in circular CFST columns than the square counterpart; whereas in the case of stiffness, it is vice versa.

Tao and Han [29] carried out a study which was almost similar to Tao et al. [28] except that rather than stub column, the researchers carried out tests on beam-columns. 28 beam-columns with 14 in circular and 14 in square cross-section were cast and divided into three categories as in the study by Tao et al. [28]. Specimens were heated in accordance with ISO-834 standard fire curve for 180 min. Repaired of fire-damaged beam-columns were done by wrapping the fire-damaged beam-columns with one or two layers of CFRP using hand lay-up method. The study showed that the ultimate strength and stiffness of fire-damaged beam-columns increased with increasing layer of CFRP. In order to quantify the enhancement in strength, the following Strength Enhancement Index (SEI) is introduced:

$$\boldsymbol{SEI} = \frac{\boldsymbol{N_{eS}} - \boldsymbol{N_{eU}}}{\boldsymbol{N_{eU}}} \tag{3.3}$$

where N_{eS} and N_{eU} is ultimate strength of repaired columns and ultimate strength of unrepaired column after exposure to fire, respectively. It was found that, the value of SEI ranged from 5.9 to 101.8% with the highest SEI value found in the case of circular beam-columns repaired with two layers of CFRP. In other words, repairing circular fire-damaged CFST beam-column with two layers of CFRP can restore the beam-column to its original ambient temperature strength.

Ductility index (DI) for stub columns from Tao et al. [28] and [29] were shown in Fig. 3.10 and Fig. 3.11, respectively. In Fig. 3.10a, it can be clearly seen that DI decreased after fire exposure. DI is also found to keep decreasing even after the fire-damaged columns were repaired with one and two layers of CFRP. However, in Fig. 3.11a, DI of circular CFST beam-columns increased after fire exposure. Similar patterns were observed in square CFST beam-columns (Fig. 3.11b). The above results meant that, the CFST beam-columns became more ductile after heating process. In addition, wrapping with CFRP enhanced the DI of both circular and square CFST beam-columns. However, the DI of circular CFST beam-column was not fully restored up to its original value.

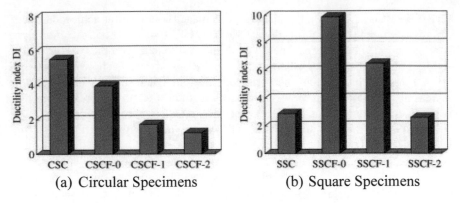

Fig. 3.10 Ductility index of concrete-filled steel tubular columns [28]

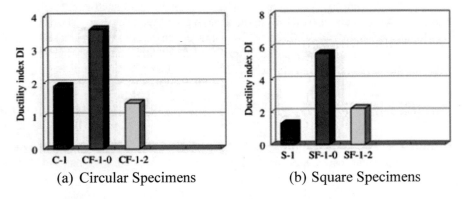

Fig. 3.11 Ductility index of concrete-filled steel tubular beam-columns [29]

References

1. Dwaikat, M.B.: Flexural Response of Reinforced Concrete Beams Exposed to Fire, pp. 1–369. Michigan State University. Doctor of Philosophy (2009)
2. Liu, L.: Fire Performance of High Strength Concrete Materials and Structural Concrete, pp. 1–229. Faculty of The College of Engineering and Computer Science. Florida Atlantic University, Florida. Doctor of Philosophy (2009)
3. Kodur, V.: Properties of concrete at elevated temperatures. Int. Sch. Res. Not. Civ. Eng. **2014**, 1–15 (2014)
4. Annerel, E., Taerwe, L.: Revealing the temperature history in concrete after fire exposure by microscopic analysis. Cem. Concr. Res. **39**(12), 1239–1249 (2009)
5. Liu, F., Gardner, L., Yang, H.: Post-fire behaviour of reinforced concrete stub columns confined by circular steel tubes. J. Constr. Steel Res. **102**, 82–103 (2014)
6. Hertz, K.D.: Concrete strength for fire safety design. Mag. Concr. Res. **57**(8), 445–453 (2005)
7. Shoukry, S.N., et al.: Effect of moisture and temperature on the mechanical properties of concrete. Constr. Build. Mater. **25**(2), 688–696 (2011)
8. Georgali, B., Tsakiridis, P.E.: Microstructure of fire-damaged concrete. A case study. Cement Concr. Compos. **27**(2), 255–259 (2005)

9. Chang, Y.F., et al.: Residual stress–strain relationship for concrete after exposure to high temperatures. Cem. Concr. Res. **36**(10), 1999–2005 (2006)
10. Chan, Y.N., Peng, G.F., Anson, M.: Residual strength and pore structure of high-strength concrete and normal strength concrete after exposure to high temperatures. Cem. Concr. Compos. **21**(1), 23–27 (1999)
11. Mohamedbhai, G.T.G.: Effect of exposure time and rates of heating and cooling on residual strength of heated concrete. Mag. Concr. Res. **38**(136), 151–158 (1986)
12. Yang, H., et al.: Evaluating residual compressive strength of concrete at elevated temperatures using ultrasonic pulse velocity. Fire Saf. J. **44**(1), 121–130 (2009)
13. Anderberg, Y.: Assessment of fire-damaged concrete structures and the corresponding repair measures. In: Concrete Repair, Rehabilitation and Retrofitting II, pp. 631–636. CRC Press (2009)
14. Qiang, X., Bijlaard, F.S.K., Kolstein, H.: Post-fire mechanical properties of high strength structural steels S460 and S690. Eng. Struct. **35**, 1–10 (2012)
15. Gunalan, S., Mahendran, M.: Thin-walled structures experimental investigation of post-fire mechanical properties of cold-formed steels. Thin Walled Struct. **84**, 241–254 (2014)
16. McCann, F., Gardner, L., Kirk, S.: Elevated temperature material properties of cold-formed steel hollow sections. Thin-Walled Struct. **90**, 84–94 (2015)
17. Moftah, M.: Numerical Modelling and Performance of Reinforced Concrete Members Under Fire Condition, pp. 1–286. Department of Civil and Environmental Engineering. Canada, The University of Western Ontario. Doctor of Philosophy (2008)
18. Lu, J., Liu, H., Chen, Z., Liao, X.: Experimental investigation into the post-fire mechanical properties of hot-rolled and cold-formed steels. J. Constr. Steel Res. **121**, 291–310 (2016)
19. Wang, Y.C.: Steel and Composite Structures. Spon Press, London (2002)
20. Ingham, J.: Forensic engineering of fire-damaged structures. Proc. ICE Civ. Eng. **162**(5), 12–17 (2009)
21. Han, L.H., Yang, H., Cheng, S.L.: Residual strength of concrete filled RHS stub columns after exposure to high temperatures. Adv. Struct. Eng. **5**(2), 123–134 (2002)
22. Imani, R., Bruneau, M., Mosqueda, G.: Simplified analytical solution for axial load capacity of concrete-filled double-skin tube (CFDST) columns subjected to fire. Eng. Struct. **102**, 156–175 (2015)
23. Lu, H., Zhao, X.L., Han, L.H.: FE modelling and fire resistance design of concrete filled double skin tubular columns. J. Constr. Steel Res. **67**(11), 1733–1748 (2011)
24. Lu, H., Han, L.H., Zhao, X.L.: Fire performance of self-consolidating concrete filled double skin steel tubular columns: experiments. Fire Saf. J. **45**(2), 106–115 (2010)
25. Lu, H., Zhao, X.L., Han, L.H.: Testing of self-consolidating concrete-filled double skin tubular stub columns exposed to fire. J. Constr. Steel Res. **66**(8–9), 1069–1080 (2010)
26. Yang, Y.F., Han, L.H.: Concrete-filled double-skin tubular columns under fire. Mag. Concr. Res. **60**(3), 211–222 (2008)
27. Yang, Y., Han, L.: Fire resistance of concrete-filled double skin steel tubular columns. Adv. Struct. Eng. **II**, 1047–1052 (2005)
28. Tao, Z., Han, L.H., Wang, L.L.: Compressive and flexural behaviour of CFRP-repaired concrete-filled steel tubes after exposure to fire. J. Constr. Steel Res. **63**(8), 1116–1126 (2007)
29. Tao, Z., Han, L.H.: Behaviour of fire-exposed concrete-filled steel tubular beam columns repaired with CFRP wraps. Thin-Walled Struct. **45**(1), 63–76 (2007)
30. Han, L., Huo, J.: Concrete-filled hollow structural steel columns after exposure to ISO-834 Fire Standard. J. Struct. Eng. **129**(1), 68–78 (2003)
31. Rush, D.I., Bisby, L.A., Jowsey, A., Lane, B.: Residual capacity of fire-exposed concrete-filled steel hollow section columns. Eng. Struct. **100**, 550–563 (2015)
32. Yaqub, M., Bailey, C.G.: Repair of fire damaged circular reinforced concrete columns with FRP composites. Constr. Build. Mater. **25**(1), 359–370 (2011)
33. Yin, J., Zha, X., Li, L.: Fire resistance of axially loaded concrete filled steel tube columns. J. Constr. Steel Res. **62**(7), 723–729 (2006)

34. Kodur, V.K.R., Mackinnon, D.H.: Design of concrete-filled hollow structural steel columns for fire endurance. Eng. J. **37**(1), 13–24 (2000)
35. Dai, X.H., Lam, D.: Shape effect on the behaviour of axially loaded concrete filled steel tubular stub columns at elevated temperature. J. Constr. Steel Res. **73**, 117–127 (2012)
36. Han, L.H., Li, W., Bjorhovde, R.: Developments and advanced applications of concrete-filled steel tubular (CFST) structures: members. J. Constr. Steel Res. **100**, 211–228 (2014)
37. Han, L.H., Zhao, X.L., et al.: Experimental study and calculation of fire resistance of concrete-filled hollow steel columns. J. Struct. Eng. **129**(3), 346–356 (2003)
38. Han, L.H., Xu, L., Zhao, X.L.: Tests and analysis on the temperature field within concrete filled steel tubes with or without protection subjected to a standard fire. Adv. Struct. Eng. **6**(2), 121–133 (2003)
39. Franssen, J., Kodur, V.K.R., Zaharia, R.: Designing Steel Structures for Fire Safety. Taylor & Francis, London (2009)
40. Tao, Z., Han, L.H., Zhuang, J.P.: Cyclic performance of fire-damaged concrete-filled steel tubular beam–columns repaired with CFRP wraps. J. Constr. Steel Res. **64**(1), 37–50 (2008)
41. Wu, Y.F., Liu, T., Oehlers, D.J.: Fundamental principles that govern retrofitting of reinforced concrete columns by steel and FRP jacketing. Adv. Struct. Eng. **9**(4), 507–533 (2006)
42. Balaguru, P., Nanni, A., Giancaspro, J.: FRP Composites for Reinforced and Prestressed Concrete Structures. Taylor & Francis, New York (2009)
43. Hu, Y.M., Yu, T., Teng, J.G.: FRP-confined circular concrete-filled thin steel tubes under axial compression. J. Compos. Constr. **15**(5), 850–860 (2011)

Chapter 4
Temperature Distribution and Post-fire Behavior of Concrete-Filled Double Skin Steel Tubular Columns

The aim of this chapter is to study the temperature distribution during fire exposure and post-fire behavior of concrete-filled double skin steel tubular (CFDST) columns after fire exposure. This chapter addresses the first objective of the study.

4.1 Physical Appearance After Fire Exposure

Generally, there is no difference in steel appearance after being exposed to both 60 and 90 min of fire. The comparison in unheated specimen and heated specimen of CFDST column is shown in Fig. 4.1. This comparison applies for both 60 and 90 min fire exposure time as well as for all Series 1, Series 2 and Series 3.

As for concrete, the color changes as the temperature increased. In addition to that, the concrete became more brittle and it crumbled under action of just a small pressure. Thermocouple reading indicates that the average maximum temperatures inside concrete were 516 °C and 563 °C for 60 and 90 min of fire exposure time, respectively. Therefore, it is expected that the concrete color changes from normal color as shown in Fig. 4.2a to whitish grey as shown in Fig. 4.2b, c. However, the difference in color is more prominent in specimens heated to 90 min. The change in color can be attributed to the transformation of ferric compounds present in aggregates or fine aggregates as impurities to ferric oxides. The intensity of color depends on the level of impurities [1]. In addition, concrete became more brittle with increase in fire exposure time. As can be seen in Fig. 4.2b, c, the concrete crumbles when the outer steel tube is removed from column. Based on previous research studies, microcracks started to form throughout the concrete when the temperature is around 300 °C due to dehydration of cement paste and thermal incompatibility between aggregates and cement paste [2, 3]. As the temperature increased (516 °C and 563 °C for 60–90 min of fire exposure time, respectively), the microcracks in concrete became more severe leading to significant lost in strength. At temperature around 400–600 °C, calcium

© The Author(s), under exclusive license to Springer Nature Singapore Pte Ltd. 2021   37
K. K. Choong et al., *Concrete-Filled Double Skin Steel Tubular Column with Hybrid Fibre Reinforced Polymer*, SpringerBriefs in Applied Sciences and Technology, https://doi.org/10.1007/978-981-16-2715-6_4

(a) Unheated specimen   (b) Heated specimen

Fig. 4.1 Physical appearance of concrete-filled double skin steel tubular columns before and after fire exposure

hydroxide ($Ca(OH)_2$) in cement breaks down into calcium oxide (CaO) and water (H_2O). In addition, aggregates also start to deteriorate at temperature around 550 °C. This leads to significant reduction in strength of concrete.

Based on previous research studies, 300 °C is marked as the critical temperature of concrete after which concrete started to lose its compressive strength. Since the maximum temperature attained by concrete is 516 and 563 °C, the concrete is expected to lose almost 85% of its original room temperature strength. However, the concrete is still able to retain its shape due to outer and inner steel tube. Concrete crumbles once the outer steel tube is removed as can be seen in Fig. 4.2.

(a) Control specimen   (b) 60 minutes fire
                                     exposure time

(b) 90 minutes fire exposure time

Fig. 4.2 Concrete condition before and after fire exposure

4.2 Effect of Different Diameter and Fire Exposure Time Towards Concrete and Inner Steel Tube Temperature

Figure 4.3a shows the influence of diameter and fire exposure time towards temperature of concrete. The temperature of concrete is almost constant for all diameters. Therefore, even with increased diameter, the maximum temperature of concrete remain almost constant for both 60 and 90 min of fire exposure time. In addition, the maximum temperature of concrete is higher for 90 min than 60 min of fire exposure time. Temperature of inner steel tube for 60 min of fire exposure time decreased

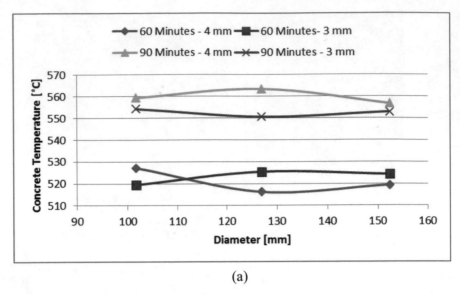

(a)

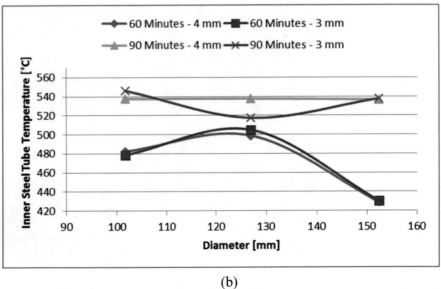

(b)

Fig. 4.3 Influence of outer steel tube diameter and fire exposure time towards **a** temperature of concrete and **b** temperature of inner steel tube

with increase diameter (Fig. 4.3b). However, this effect seems to diminish for 90 min fire exposure time. The increment in diameter leads to higher concrete volume in the specimen which acts as heat sink thus protecting inner steel tube. As the fire exposure time increased, concrete absorbs more heat which causes rise in the inner steel tube temperature consequently. Nevertheless, the critical temperature of steel for inner steel tube was never reached. This can also be related to UBBL (1984) requirement, which state that the concrete thickness of 50.8 mm is able to withstand up to 2 h of fire. This is proven to be the case for specimens in this research study where the concrete thickness of all the Series is constant i.e., 50.8 mm.

4.2.1 Time-Temperature Curve of Concrete-Filled Double Skin Steel Tubular (CFDST) Columns

During fire test, the temperature of furnace, concrete and inner steel tubes were recorded. However, for every fire exposure time (i.e., 60, 90 min) and thickness, only three specimens were equipped with thermocouple inside the concrete. The time-temperature curve for all specimens (except the one with damaged concrete thermocouple) is shown in Fig. 4.4. The furnace temperature is an average reading of four thermocouple attached to the wall of the furnace. The furnace temperature is in accordance with ASTM E-119 (2010) standard fire curve. However, the time-temperature curve is quite unstable. Clause 7.2.3 in ASTM E-119 (2010) states that the difference in area under time-temperature curve between ASTM E-119 (2010) time-temperature curve and recorded time-temperature curve should follow the guidelines as shown in Table 4.1. Based on Table 4.1, the difference in area under the time-temperature curve for 60 min of fire exposure time should be less or equal to 10 and 7.5% for 90 min of fire exposure time. The values for all specimens are calculated and compared against this clause. Table B.3 in Appendix B listed the calculated value for all heated specimens. The calculation of area under the curve is done by using trapezoidal method. The curve is divided into small trapezoid as shown in Fig. 4.5. The area of each small trapezoid is marked as A_i. Total area under the time-temperature curve is calculated by adding all A_i. It can be seen that all curves are within the acceptable limit.

There is relatively flat plateau for time-temperature curves for concrete as shown in Fig. 4.4a. This happens when the temperature is around 100–200 °C. During this time, the heat is used to vaporize water inside the concrete rather than raising the temperature of concrete and inner steel tube. This phenomenon is also observed by other researchers [4, 5]. The evaporation of free moisture in the concrete started to occur when temperature reached 100 °C, hence the flat plateau. During this stage, if concrete is not confined by steel tube spalling is most likely to occur due to build up pressure inside concrete. However, spalling is not the main concern in CFDST columns [6]. At around 250 °C, the evaporation of capillary water begins and ended when temperature reached 400 °C [7].

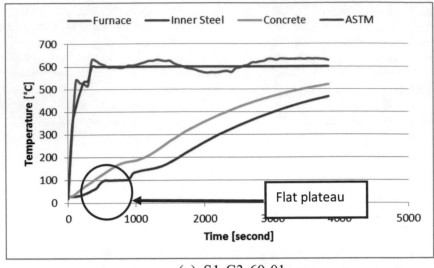

(a) S1-C3-60-01

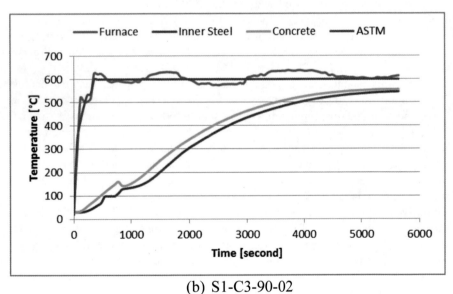

(b) S1-C3-90-02

Fig. 4.4 Time-temperature curve of concrete-filled double skin steel tubular columns during fire exposure

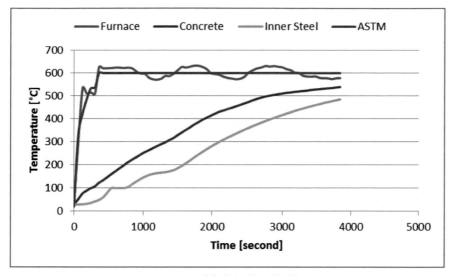

(c) S1-C4-60-02

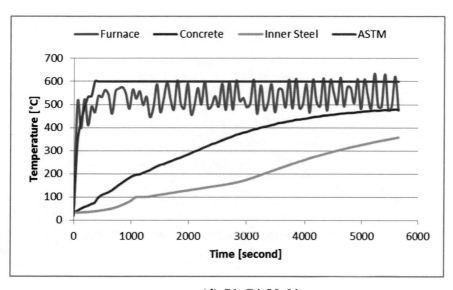

(d) S1-C4-90-01

Fig. 4.4 (continued)

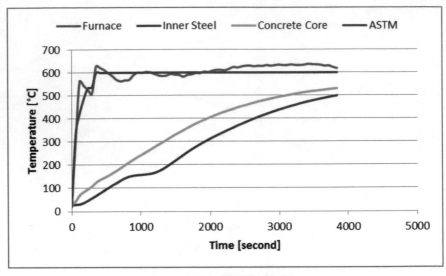

(e) S2-C3-60-01

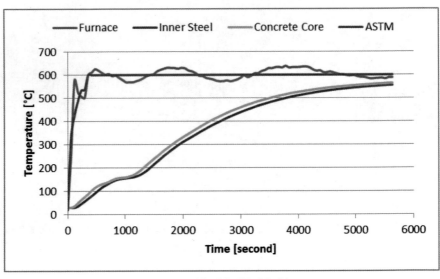

(f) S2-C3-90-01

Fig. 4.4 (continued)

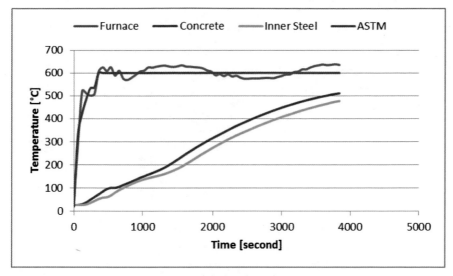

(g) S2-C4-60-01

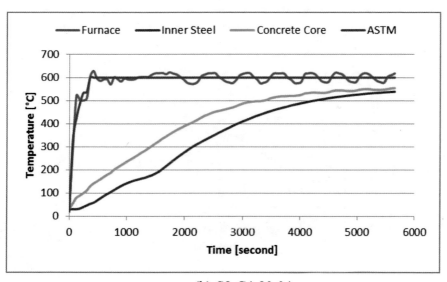

(h) S2-C4-90-04

Fig. 4.4 (continued)

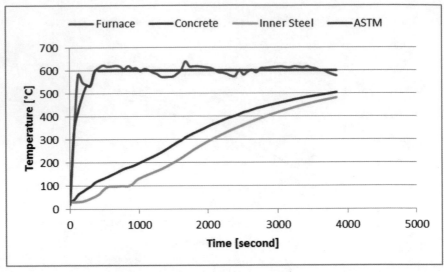

(i) S3-C3-60-01

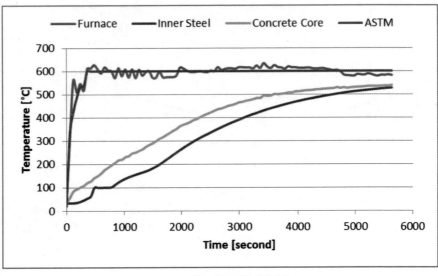

(j) S3-C3-90-01

Fig. 4.4   (continued)

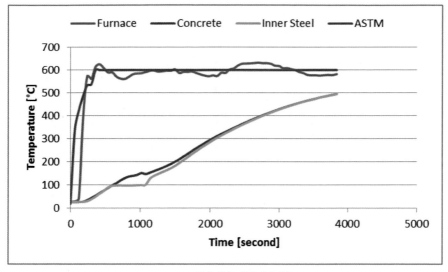

(k) S3-C4-60-01

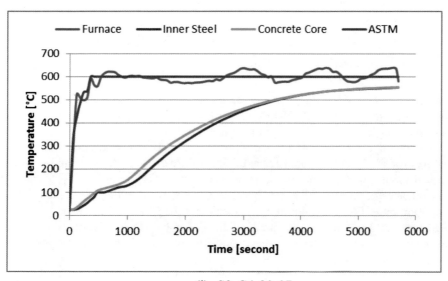

(l) S3-C4-90-07

Fig. 4.4 (continued)

Table 4.1 Tolerances for
time-temperature curve
(ASTM 2010)

Area under time-temperature curve (%)	Exposure time, t
10.0	$t \leq 1\,h$
7.5	$1\,h < t \leq 2\,h$
5.0	$t > 2\,h$

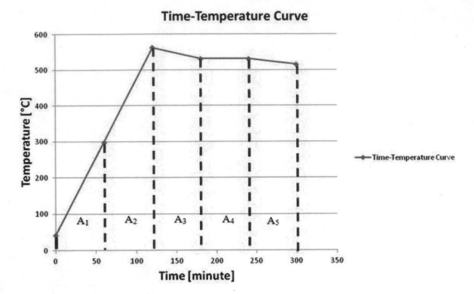

Fig. 4.5 Trapezoidal method

4.3 Effect of Thickness on Residual Strength Index, Secant Stiffness and Ductility Index

Effects of thickness on RSI, secant stiffness and DI of fire exposed concrete-filled double skin steel tubular (CFDST) columns after 60 and 90 min of fire exposure time are shown in Fig. 4.6, Fig. 4.7 and Fig. 4.8, respectively.

In general, even with only 1 mm difference, t_o affects RSI, secant stiffness and DI of fire exposed CFDST columns. In both cases of fire exposure time (60 and 90 min), $t_o = 4$ mm is found to perform better than $t_o = 3$ mm. It is interesting to note that the performance of thicker t_o is more obvious when subjected to longer duration of fire exposure time. From Fig. 4.6, it can be seen that for 60 min of fire exposure time, RSI for $t_o = 3$ mm and $t_o = 4$ mm is relatively similar. However for 90 min of fire exposure time, the difference between $t_o = 3$ mm and $t_o = 4$ mm is more pronounce. RSI for $t_o = 4$ mm is lower than RSI for $t_o = 3$ mm under 90 min of fire exposure time. This difference in effect of RSI means that, CFDST columns need thicker t_o in order to minimize damaged done by fire. With only 1 mm difference in thickness, longer fire resistance of CFDST columns can be achieved.

Secant stiffness of all fire exposed CFDST columns increased with increase t_o. The secant stiffness of $t_o = 4$ mm is higher than that of $t_o = 3$ mm for both 60 and 90 min of fire exposure time (Fig. 4.7). This phenomenon is as expected since thicker t_o results in stiffer columns. In addition to that, thicker t_o can delay the temperature rises in CFDST columns thus preserving the stiffness of the columns.

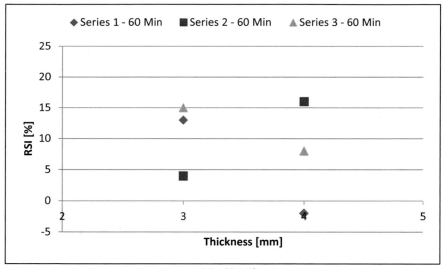

(a) 60 Minutes

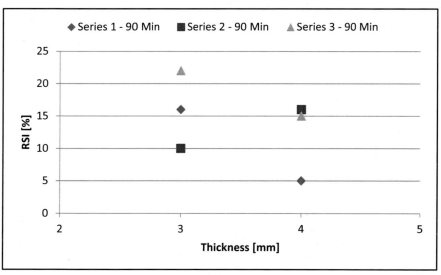

(b) 90 Minutes

Fig. 4.6 Effect of thickness on residual strength of fire exposed concrete-filled double skin steel tubular columns

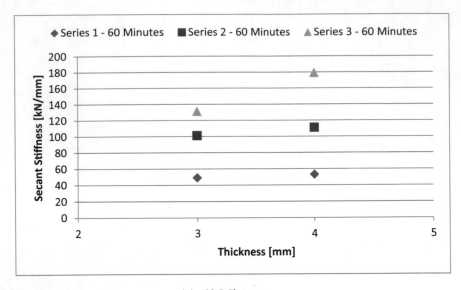

(a) 60 Minutes

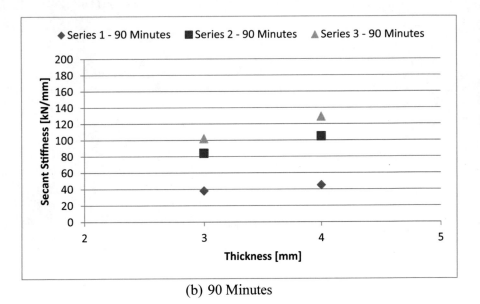

(b) 90 Minutes

Fig. 4.7 Effect of thickness on secant stiffness of fire exposed concrete double skin steel tubular columns

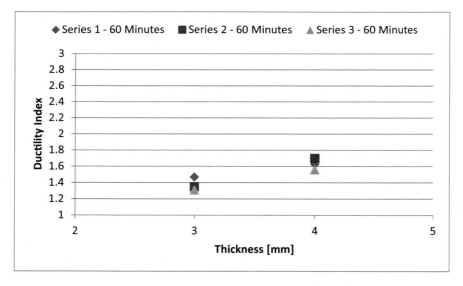

(a) 60 Minutes

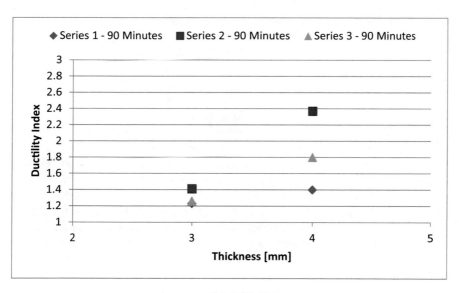

(b) 90 Minutes

Fig. 4.8 Effect of thickness on ductility index of fire exposed concrete filled double skin steel tubular columns

After being exposed to 90 min of fire exposure time, CFDST column with $t_o =$ 4 mm is found to show higher value of DI than its $t_o = 3$ mm counterpart. This is shown in Fig. 4.8. In other words, after being exposed to 90 min of fire exposure time, CFDST columns with $t_o = 4$ mm become more ductile. The increment in DI after being exposed to fire is also observed by Liu et al. [8] for concrete-filled steel tube (CFST) column under 30 and 60 min of fire exposure time. Liu et al. [8] used 250 mm diameter of outer steel tube with $t_o = 2.20$ mm and $t_o = 1.76$ mm. Therefore, it can be concluded that in term of DI, the behavior of CFDST columns are similar to that of CFST columns after exposure to fire.

4.4   Effect of Fire Exposure Time on Residual Strength Index, Secant Stiffness and Ductility Index

RSI increased with increase in fire exposure time regardless of the diameter of specimen. Longer fire exposure time increased the maximum temperature attained by all of the elements of the CFDST column (i.e., outer steel tube, concrete and inner steel tube) thus reducing the ultimate load of the specimen. However, as pointed out earlier, RSI for S1-C4-60 is negative, which means that instead of losing strength after being exposed to fire, the specimen in this Series gains strength by 2%. Moreover, the highest value of RSI after 90 min of fire exposure time is 22% (S3-C3-90). In other words, the specimen only loses 22% of its control strength after being exposed to fire for 90 min.

Secant stiffness of fire exposed CFDST columns decreased with increase in exposure time. Specimens with 90 min of fire exposure time absorbed more heat than specimens with 60 min of fire exposure time hence affecting secant stiffness of the specimen. In addition, the difference in secant stiffness is more obvious as the diameter is increased where the highest value was observed in S3-C4-60; whereas the lowest value is observed in specimen S3-C3-90. Larger diameter contains more concrete mass thus the concrete is able to protect inner steel tube when the specimen is exposed for 60 min of fire. However, 90 min of fire exposure time with thinner t_0 are unable to protect inner steel tube thus lead to lowest secant stiffness in all Series. Apart from that, both C3-60 and C4-90 in all three Series show similar increment in secant stiffness as diameter increases from Series 1 to Series 3. This means that, with only 1 mm difference in t_0, the specimen can withstand additional 30 min of fire exposure time while maintaining similar secant stiffness.

Effect of fire exposure time on DI is dependent on t_0 and diameter of the specimen as shown in Fig. 4.9. For Series 1 (diameter = 101.6 mm), DI is found to decrease with increase in fire exposure time and decrease in t_0. However, as the diameter is increased, DI is found to increase as exposure time and t_0 increase. For thinner t_0 ($t_0 = 3$ mm), DI decreased with increase in fire exposure time. In addition, with increased diameter, exposure time and t_0, DI for fire exposed CFDST column is found to exceed the DI of control specimen. Increase in DI under longer fire exposure time

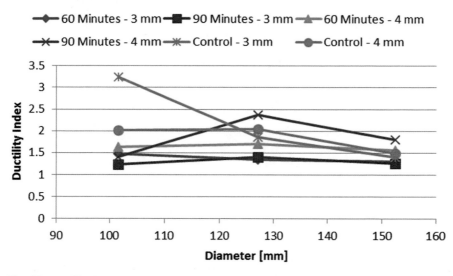

Fig. 4.9 Ductility index of concrete-filled double skin steel tubular columns

can be caused by more heat being absorbed by the specimen thus weakened the specimen and affected the ability of the specimen to carry load.

In Series 1, DI for both 60 and 90 min of fire exposure time is lower than DI of control specimen (Fig. 4.9). In this Series, the specimens become less ductile after fire. However, as the diameter increased from 101.6 mm (Series 1) to 127 mm (Series 2), the specimen with 90 min of fire exposure time and $t_0 = 4$ mm become more ductile than the other specimens. DI of this specimen exceeded the DI of control specimen. In Series 3, specimen with 60 and 90 min of fire exposure time and $t_0 = 4$ mm is observed to have an increase in DI exceeding the DI of control specimen.

References

1. Yaqub, M., Bukhari, I.A., Ghani, U.: Assessment of residual strength based on estimated temperature of post-heated RC columns. Mehran Univ. Res. J. Eng. Technol. **32**(1), 55–70 (2013)
2. Chan, Y.N., Peng, G.F., Anson, M.: Residual strength and pore structure of high-strength concrete and normal strength concrete after exposure to high temperatures. Cement Concr. Compos. **21**(1), 23–27 (1999)
3. Hertz, K.D.: Concrete strength for fire safety design. Mag. Concr. Res. **57**(8), 445–453 (2005)
4. Lu, H., Han, L.H., Zhao, X.L.: Fire performance of self-consolidating concrete filled double skin steel tubular columns: experiments. Fire Saf. J. **45**(2), 106–115 (2010)
5. Lu, H., Zhao, X.L., Han, L.H.: Testing of self-consolidating concrete-filled double skin tubular stub columns exposed to fire. J. Constr. Steel Res. **66**(8–9), 1069–1080 (2010)
6. Georgali, B., Tsakiridis, P.E.: Microstructure of fire-damaged concrete. A case study. Cement Concr. Compos. **27**(2), 255–259 (2005)

7. McCann, F., Gardner, L., Kirk, S.: Elevated temperature material properties of cold-formed steel hollow sections. Thin-Walled Struct. **90**, 84–94 (2015)
8. Liu, F., Gardner, L., Yang, H.: Post-fire behaviour of reinforced concrete stub columns confined by circular steel tubes. J. Constr. Steel Res. **102**, 82–103 (2014)

Chapter 5
Repair of Fire-Damaged Concrete-Filled Double Skin Steel Tubular Columns with Fiber Reinforced Polymer (FRP)

This chapter is focused on the repair of fire-damaged concrete-filled double skin steel tubular (CFDST) columns with two different repairing schemes. The first scheme is by using one layer of Carbon Fiber Reinforced Polymer (CFRP) and the second one with two layers of Fiber Reinforced Polymer (FRP). The latter scheme is called Hybrid Repair in this thesis, where the first layer consists of Glass Fiber Reinforced Polymer (GFRP) and the second layer CFRP. The performances of repaired CFDST columns are discussed in detail.

5.1 Failure Pattern of Repaired Fire-Damaged Concrete-Filled Double Skin Steel Tubular (CFDST) Column

Repaired of fire-damaged concrete-filled double skin steel tubular (CFDST) columns failed by outwards local buckling of outer steel tube, crushing of concrete and local buckling of inner steel tube, which is similar to that of control and unrepaired specimens. However, local buckling of outer steel tube started after failure of FRP confining the CFDST columns. FRP failed by rupture in the hoop direction as shown in Figs. 5.1 and 5.2.

Prior to rupture of FRP, cracking sound can be heard indicating full activation of FRP. The rupture of FRP is sudden and quite explosive. However, it is more gradual for two layer of FRP (Hybrid FRP). This can be attributed to the presence of Glass FRP (GFRP). GFRP possess higher ultimate tensile strain than CFRP, thus it can expand more than CFRP before rupture. On the contrarily, Carbon FRP (CFRP) is more brittle in nature when compared to GFRP. Thus, the failure of CFRP is more violent than GFRP. The sudden and explosive failure of FRP is due to the release of a massive amount of energy from the confining stress provided by the FRP. In addition,

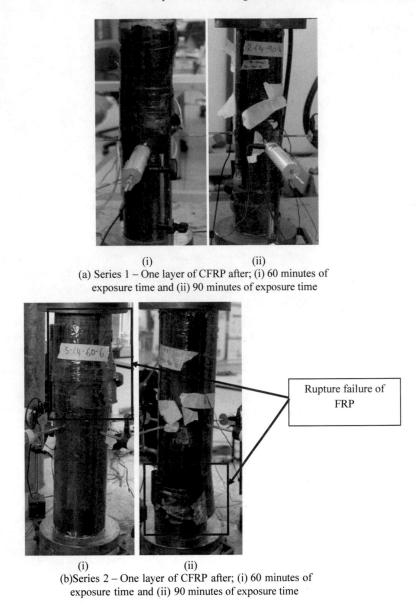

(i)   (ii)
(a) Series 1 – One layer of CFRP after; (i) 60 minutes of
exposure time and (ii) 90 minutes of exposure time

Rupture failure of
FRP

(i)   (ii)
(b)Series 2 – One layer of CFRP after; (i) 60 minutes of
exposure time and (ii) 90 minutes of exposure time

Fig. 5.1 Failure pattern of repaired fire-damaged concrete-filled double skin steel tubular columns

the rupture of FRP occurred either at the top or bottom ends of the specimens and
never at the overlap position as shown in Fig. 5.2. For specimens wrapped with one
layer of CFRP, the rupture started at the top or bottom ends. It is then propagated
until mid height of the specimens. Nevertheless, rupture of FRP only occurred either
at the top or bottom end for specimens repaired with Hybrid FRP. Since the loading

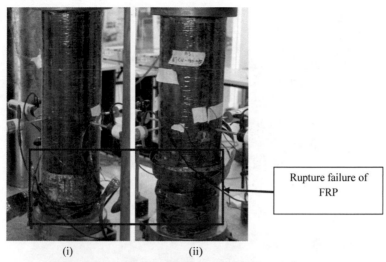

(i)   (ii)

(c)Series 3 – One layer of CFRP after; (i) 60 minutes of
exposure time and (ii) 90 minutes of exposure time

(i)   (ii)

(d)Series 3 –Two layers of GFRP and CFRP after; (i) 60 minutes of
exposure time and (ii) 90 minutes of exposure time

Fig. 5.1 (continued)

Fig. 5.2 Rupture of fiber reinforced polymer

is stopped after the load dropped to around 80% of the maximum load, the rupture of FRP for the specimen with Hybrid FRP only occurred either at the top or bottom end of the specimen (Fig. 5.2).

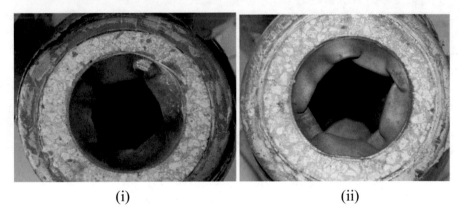

(i)   (ii)

Fig. 5.3 Failure pattern of inner steel tube of concrete-filled double skin steel tubular columns; (i) 60 min of fire exposure time and (ii) 90 min of fire exposure time

5.2 Repaired of Fire-Damaged Concrete-Filled Double Skin Steel Tubular (CFDST) Columns with Single Layer of Fiber Reinforced Polymer (FRP)

5.2.1 Effect of Fiber Reinforced Polymer (FRP) Wrap on Ultimate Load

The results shown in Table 5.1 are the average values from three specimens. Additional letter R and RH at the end of specimen naming convention refers to specimens repaired with one layer of FRP and specimens repaired with Hybrid FRP, respectively. Ultimate load for control specimens, fire exposed specimens and repaired specimens are compared in Fig. 5.4. From the figure, it can be seen that ultimate load increased after the specimens are repaired by one layer of CFRP. Apart from specimen S2-C4-60 (Fig. 5.4b), other specimen repaired with one layer of CFRP is found to have an ultimate load exceeding the ultimate load of control specimen. However, in the case of S2-C4-60, the difference in ultimate load of S2-C4-60 and S2-C4-Control (Table 5.1) is only 40 kN. This means, the gain in ultimate load after being repaired with one layer of CFRP is 96% of the ultimate load of control specimen. Even though, the ultimate load of S2-C4-60 after repair is not exceeding the ultimate load of control specimen, wrapping the fire-damaged specimen with only one layer of CFRP is able to restore 96% of its control ultimate load.

For specimens under 60 min of fire exposure time, the Strength Enhancement Index, SEI is found to range from 1% (S1-C4-60) to 32% (S2-C3-60). As for those under 90 min of fire exposure time, the lowest and highest SEI is from 8% to 26% for S1-C4-90 and S2-C4-90, respectively. It can be seen that the highest SEI is from specimen under 60 min of fire exposure time since the damaged caused by fire is less severe than 90 min of fire exposure time. Thus the enhancement in strength after repaired is greater. Therefore, shorter fire exposure time leads to higher SEI.

Effect of thickness on SEI of repaired specimens is dependent on its fire exposure time. When the fire exposure time is 60 min, SEI decreased as the t_0 increased as can be seen in Fig. 5.5a. However, increased fire exposure time seem to yield different result. Specimen with thinner t_0 show small increment in SEI. Nevertheless, specimen with thicker t_0 show wide range of SEI. This might be caused by the degree of damaged caused by fire for specimens with longer fire exposure time and thinner t_0. As discussed in the previous chapter, specimens with thinner t_0 suffered severely in the case of longer fire exposure time. However, specimens with $t_o = 4$ mm (90 min of fire exposure time) are able to maintain low RSI thus resulted in higher SEI. For specimens with 60 min of fire exposure time, CFRP is found to be effective in enhancing the ultimate load of specimen with thinner t_0. Thinner t_0 contribution to load carrying capacity is smaller than thicker t_0, thus increased the effectiveness of FRP wrap. Moreover, under axial loading, specimen with $t_o = 3$ mm is able to expand much easier than those with $t_o = 4$ mm. As a result, confinement provided

Table 5.1 Result of tested concrete-filled double skin steel tubular columns

Diameter of outer steel tube (mm)	Specimens	Ultimate load (kN)	Displacement at ultimate load (mm)	Residual strength index (%)	Strength enhancement index (%)	Secant stiffness (kN/mm)	Ductility index
101.6	S1-C3-Control	963	30.05	0	–	32	1.08
	S1-C3-60	841	17.15	13	–	49	1.47
	S1-C3-60-R	977	10.48	−1	16	93	1.53
	S1-C3-90	811	21.24	16	–	38	1.23
	S1-C3-90-R	948	14.12	2	17	67	1.35
127.0	S2-C3-Control	734	5.17	0	–	142	1.86
	S2-C3-60	703	6.96	4	–	101	1.35
	S2-C3-60-R	926	11.37	−26	32	81	1.09
	S2-C3-90	662	7.92	10	–	84	1.41
	S2-C3-90-R	795	10.64	−8	20	75	1.06
152.4	S3-C3-Control	1125	4.48	0	–	251	1.40
	S3-C3-60	951	7.29	15	–	131	1.31
	S3-C3-60-R	1124	10.10	0	18	111	1.14
	S3-C3-60-RH	1157	10.24	−3	22	113	1.13
	S3-C3-90	881	8.64	22	–	102	1.26
	S3-C3-90-R	996	10.28	11	13	97	1.13
	S3-C3-90-RH	1098	10.48	2	25	105	1.20
101.6	S1-C4-Control	996	9.27	0	–	108	2.02
	S1-C4-60	1015	19.29	−2	–	53	1.63
	S1-C4-60-R	1021	16.16	−3	1	63	1.57
	S1-C4-90	944	24.10	5	–	39	1.40
	S1-C4-90-R	1015	14.89	−2	8	68	2.01
127.0	S2-C4-Control	938	4.11	0	–	228	2.04
	S2-C4-60	792	7.14	16	–	111	1.70
	S2-C4-60-R	898	12.14	4	14	74	2.34
	S2-C4-90	790	7.51	16	–	105	2.37
	S2-C4-90-R	998	10.69	−6	26	93	1.25
152.4	S3-C4-Control	1402	6.98	0	–	201	1.49
	S3-C4-60	1292	7.22	8	–	179	1.56
	S3-C4-60-R	1428	11.69	−2	11	122	1.17
	S3-C4-60-RH	1494	11.57	−7	16	129	1.18
	S3-C4-90	1199	9.26	15	–	129	1.80
	S3-C4-90-R	1401	12.39	0	17	113	1.27
	S3-C4-90-RH	1386	10.90	1	16	127	1.18

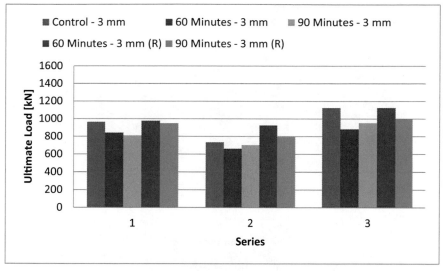

(a) $t_o = 3\ mm$

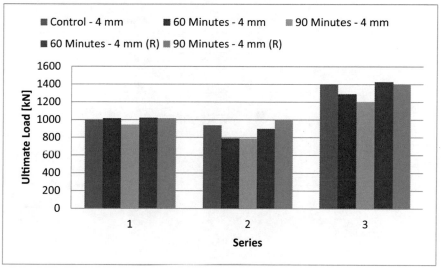

(b) $t_o = 4\ mm$

Fig. 5.4 Ultimate load of repaired fire-damaged concrete-filled double skin steel tubular columns with single layer of CFRP

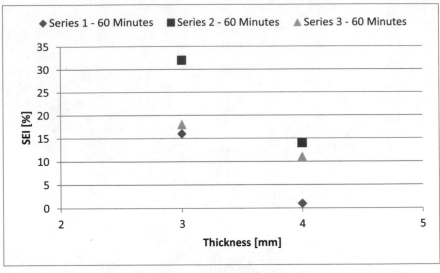

(a) 60 Minutes

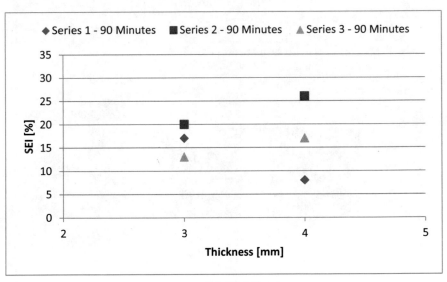

(b) 90 Minutes

Fig. 5.5 Effect of thickness on strength enhancement index of repaired concrete-filled double skin steel tubular columns

by FRP for specimens with $t_o = 3$ mm can be fully activated as reflected by higher SEI value.

From Fig. 5.5, it can be seen that regardless of t_0 and fire exposure time, Series 1 shows the lowest SEI except for specimens under 90 min of fire exposure time with $t_0 = 3$ mm. This is then followed by Series 3. The highest SEI is observed in specimen from Series 2. Therefore, it can be concluded that one layer of CFRP can effectively enhance the strength of specimens in Series 2. For fire-damaged specimens with larger diameter, CFRP is still effective in enhancing the strength. However, in order to restore the strength to its original or exceeding the original strength of the control specimen, larger diameter specimens needed to be exposed for no more than 60 min of fire exposure time. Combination of 60 min of fire exposure time with thinner t_0 resulted in highest SEI as can be clearly seen in Fig. 5.5a.

5.2.2 Effect of Fiber Reinforced Polymer (FRP) Wrap on Secant Stiffness

Secant stiffness of Series 1 increased after being repaired with one layer of CFRP (Fig. 5.6). The increment is larger for Series 1 under 60 min of fire exposure time and $t_0 = 3$ mm. This is another proof that FRP efficiently enhanced secant stiffness of CFDST specimens with thinner t_0. Under longer fire exposure time for Series 1, the increment in secant stiffness is found to be similar for both cases of t_0. However, the enhancement of secant stiffness of repaired specimens (for Series 1) is not fully restored up to the secant stiffness of control specimen.

On the other hand, repairing the fire damaged specimens of Series 2 and Series 3 with one layer of CFRP is unable to restore the secant stiffness. No significant enhancement in secant stiffness can be observed even though the CFRP greatly enhanced the strength. This can be seen from the axial load-axial strain graph in Figs. 5.7, 5.8 and 5.9. The gradient of first portion of graph represents stiffness of the specimens. In addition, the secant stiffness is further reduced after repair. This is especially true for specimens under 60 min of fire exposure time. For specimens with $t_0 = 3$ mm, the reductions were 20 kN/mm (19.8%) from fire damaged specimens for both Series. However, the reduction became larger for $t_0 = 4$ mm where it is 37 kN/mm (33.3%) and 57 kN/mm (31.8%) for Series 2 and Series 3, respectively. Nevertheless, the reduction ranges from 5 kN/mm (4.9%) (S3-C3-90) to 16 kN/mm (12.4%) (S3-C4-90) for 90 min of fire exposure time. Therefore, repairing the fire-damaged specimens with one layer of CFRP is unable to restore the secant stiffness of the specimens.

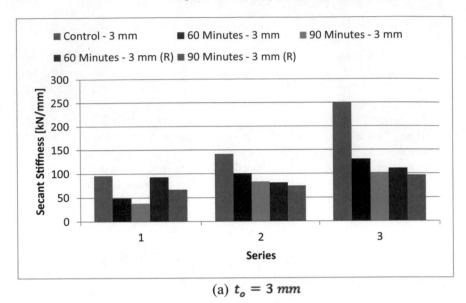

(a) $t_o = 3\ mm$

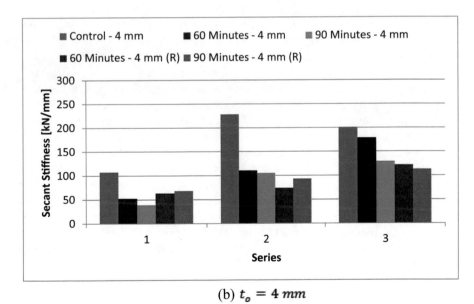

(b) $t_o = 4\ mm$

Fig. 5.6 Secant stiffness of repaired fire-damaged concrete-filled double skin steel tubular columns with single layer of CFRP

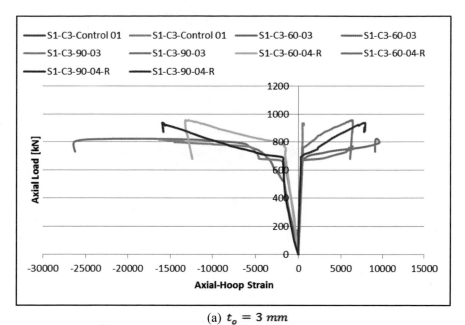

(a) $t_o = 3\ mm$

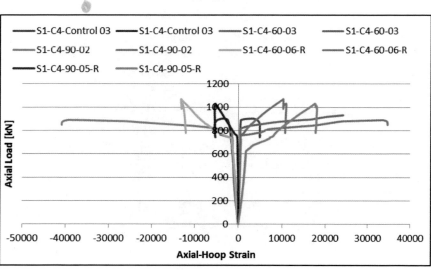

(b) $t_o = 4\ mm$

Fig. 5.7 Axial load versus axial-hoop strain curves of fire-damaged and repaired CFDST columns (Series 1, diameter of outer steel tube = 101.6 mm)

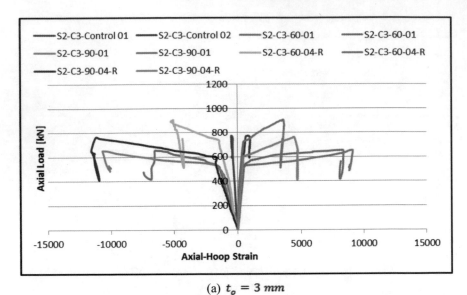

(a) $t_o = 3\ mm$

(b) $t_o = 4\ mm$

Fig. 5.8 Axial load versus axial-hoop strain curves of fire-damaged and repaired CFDST columns (Series 2, diameter of outer steel tube = 127.0 mm)

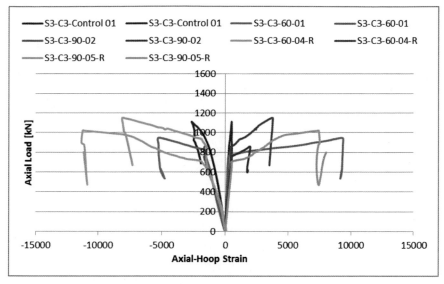

(a) $t_o = 3\ mm$

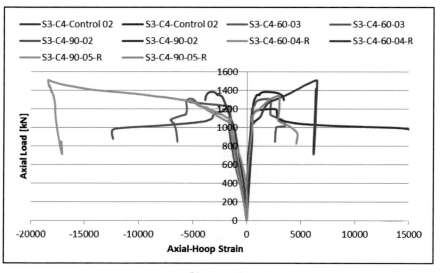

(b) $t_o = 4\ mm$

Fig. 5.9 Axial load versus axial-hoop strain curves of fire-damaged and repaired CFDST columns (Series 3, diameter of outer steel tube = 152.4 mm)

5.2.3 Effect of Fiber Reinforced Polymer (FRP) Wrap on Ductility Index (DI)

Similar to secant stiffness, DI of repaired specimens only showed an increment when the diameter is small (Series 1). DI of control, fire-damaged and repaired specimens is shown in Fig. 5.10. Series 1 with $t_o = 3$ mm experienced increment in DI for both cases of 60 and 90 min of fire exposure time. However, it is still unable to restore the DI up to the DI of control specimens. As for Series 1 with $t_o = 4$ mm, there is an increase in DI for repaired specimen under 90 min of fire exposure time. The enhancement in DI is nearly comparable to the DI of control specimen. On the contrary, specimens under 60 min of fire exposure time experienced reduction in DI. In addition, there is an increment in DI for one of the Series 2 repaired specimen, i.e., S2-C4-60-R (Fig. 5.10b). Unexpectedly, the increment in DI is found to exceed the DI of control specimen. The increment in DI is 12.8% more than the DI of control specimen. However, the increment in DI that exceeded control specimen DI only occurred in this particular specimen and none on the other repaired specimen. Therefore, it can be concluded that as the diameter increased, DI of repaired specimen is observed to experience reduction rather than increment. However, the reduction in DI is smaller for $t_0 = 3$ mm. Thus, repair using FRP performs better for specimen with thinner t_0.

5.3 Repair of Fire-Damaged Concrete-Filled Double Skin Steel Tubular (CFDST) Column with Hybrid Fiber Reinforced Polymer (FRP)

Figures 5.11 and 5.12 show the ultimate load and Strength Enhancement Index (SEI) of fire damaged specimens repaired with single layer of CFRP and Hybrid FRP. Both ultimate load and SEI increased as layer of FRP increased from single to Hybrid FRP. Similar pattern of enhancement can be seen in Fig. 5.12. The SEI is as low as 11% and as high as 25% for S3-C4-60 and S3-C3-90, respectively. Overall, wrapping fire damaged specimens with single and Hybrid FRP is proven to be able to restore the ultimate load. The enhancement in both ultimate load and SEI is higher for specimen repaired with Hybrid FRP than those repaired with single FRP due to double confinement provided by Hybrid FRP.

From Fig. 5.12, the highest SEI for specimen repair with Hybrid FRP is observed in specimens with $t_o = 3$ mm for both 60 and 90 min of fire exposure time. Therefore, Hybrid FRP is also proven to be able to effectively confine thinner t_0 than thicker t_0. For $t_o = 3$ mm, the increment of SEI for 60 min and 90 min of fire exposure time is 4% and 12%, respectively from single to Hybrid layer of FRP; whereas for $t_o = 4$ mm, the increment in SEI from single to Hybrid FRP for 60 min of fire exposure time is 5%. However, for 90 min of fire exposure time, there is no significant difference when the

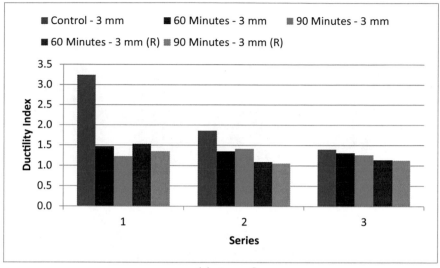

(a) $t_o = 3\ mm$

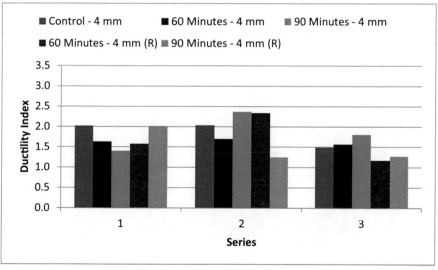

(b) $t_o = 4\ mm$

Fig. 5.10 Ductility Index of repaired fire-damaged concrete-filled double skin steel tubular columns with single layer of CFRP

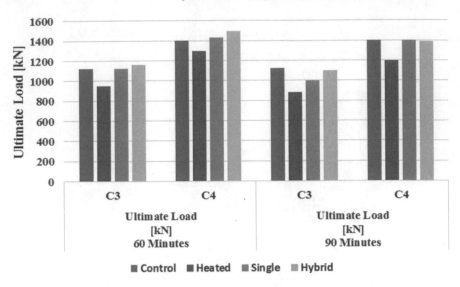

Fig. 5.11   Effects of single and hybrid layer of FRP wrapped on ultimate load

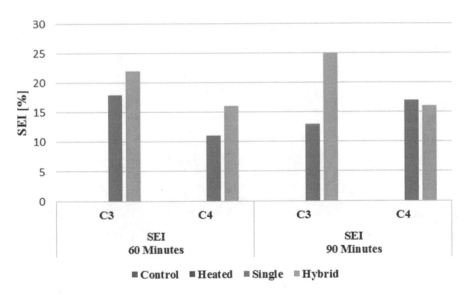

Fig. 5.12   strength enhancement index of repaired fire-damaged concrete-filled double skin steel tubular columns

fire-damaged specimens are repaired with single and Hybrid FRP (Fig. 5.12). This proves that the effectiveness of repair using FRP is affected by the t_0.

Overall, repairing of fire-damaged specimens with Hybrid layer of FRP resulted in strength enhancement exceeding the strength of control specimen. However, t_0 as well as fire exposure time of the specimens needs to be considered. If the specimens are exposed to fire for less than 60 min and the t_0 is thin, the use of single layer FRP is able to restore the strength of the specimen up to the strength of control specimen. Nevertheless, if the specimens are exposed to more than 60 min of fire, Hybrid FRP is the most suitable repairing scheme that should be adopted.

The secant stiffness of specimen repaired with FRP increased with increase in layer of FRP. This is shown in Figs. 5.13 and 5.14. The increment in secant stiffness is observed in both cases of t_0 ($t_0 = 3$ mm and $t_0 = 4$ mm) and fire exposure time (60 and 90 min). However, the increment in secant stiffness is still not sufficient to fully restore the secant stiffness up to secant stiffness of control specimen. Nevertheless, it is expected that increasing the number of FRP layers will increase the confinement, thus increasing the secant stiffness of the fire-damaged specimens.

DI of repaired CFDST columns with single and Hybrid FRP are shown together in Fig. 5.15. The DI of specimen repaired with Hybrid FRP remains almost constant in comparison with specimens repaired with single layer of CFRP which could be attributed to failure mode (sudden rupture of FRP) of the FRP. The FRP is unable to provide lateral restraint to the deformation of the specimen once it failed. The more layers of FRP resulted in more sudden the nature of failure leading to almost constant DI. Sudden rupture of multilayers of FRP is also observed by other researchers such as Tao et al. [1] and Che et al. [2]. However, both researchers used only one type of FRP in their research study, namely CFRP. Therefore, repairing fire damaged specimen with Hybrid FRP yielded similar result in DI as that using single layer FRP.

Axial strain and hoop strain in Figs. 5.7, 5.8 and 5.9 is taken directly from strain gauges attached at the mid-height of the tested specimens. Initially, it can be seen that

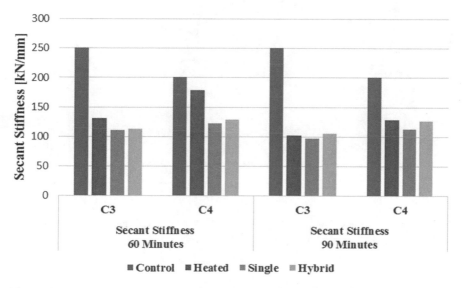

Fig. 5.13 Secant stiffness of repaired fire-damaged concrete-filled double skin steel tubular columns with Hybrid FRP

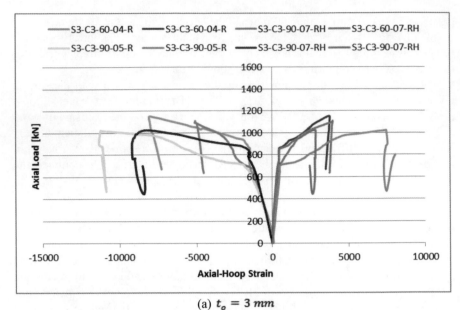

(a) $t_o = 3\ mm$

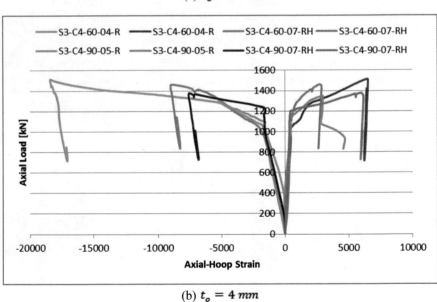

(b) $t_o = 4\ mm$

Fig. 5.14 Axial load-hoop strain curves of fire-damaged and repaired concrete-filled double skin steel tubular columns with single and Hybrid FRP

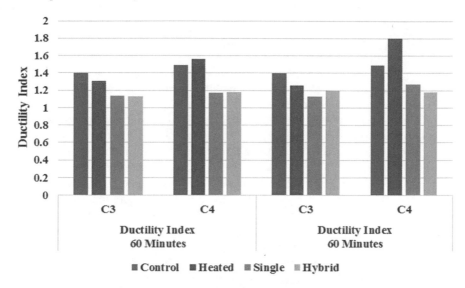

Fig. 5.15 Ductility index of repaired fire-damaged concrete-filled double skin steel tubular columns with hybrid FRP

the axial load-hoop strain curves for all specimens follow similar trend. There is small expansion in hoop direction until the specimen reaches unconfined strength of the specimen (Fig. 5.16). This expansion occurred in control and heated specimens. The

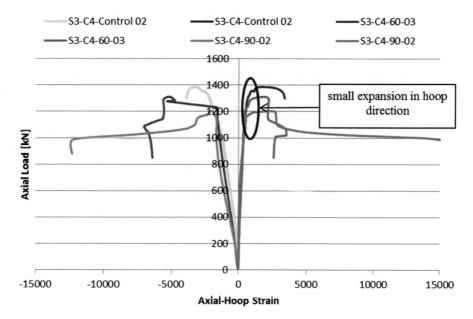

Fig. 5.16 Axial load versus axial-hoop strain curves of Series 3 with $t_o = 4$ mm

expansion is due to the expansion of steel and concrete under increased axial load. Soon after this point, there is significant increase in hoop strain. With increased hoop strain, both control and heated specimen experienced gradual decrement or in some cases constant axial load. This can be attributed to the occurrence of local buckling of the specimens. This is also associated with the steel tube losing its confinement ability due to expansion of concrete.

Axial load-hoop strain curves of repaired specimens have no expansion in hoop direction as in control and heated specimens (Fig. 5.17). After the first portion of graph, both axial load and hoop strain increased steadily as label in Fig. 5.17. During this stage, the CFRP effectively confined the steel tube. The presence of FRP delayed the occurrence of local buckling due to the confinement provided by the FRP wrapped. In addition, the specimens are able to carry more loads until failure of FRP. The failure of FRP is associated with sudden drop in axial load. During this stage, the additional strength provided by FRP confinement is no longer available thus resulting in sudden drop in load.

Based on the observation and from the axial load-hoop strain curves, wrapping with single layer of FRP on fire-damaged specimen is proven to be able to delay the occurrence of local buckling. Besides, the confinement provided by FRP further improved the load carrying capacity of fire-damaged specimens.

Axial load versus axial-hoop strain of fire-damaged specimens repaired with single and Hybrid layer of FRP is shown in Fig. 5.18. The curves for both specimens repaired with single layer and Hybrid FRP are similar. It consists of rapid linear ascending of the first portion of the curve. Then, it is followed by gradual increased in hoop strain which indicates confinement provided by FRP. Finally, it ended with sudden dropped in axial load due to failure of FRP. In addition, the curves for Hybrid FRP are steeper than single FRP. This indicates that, under similar load, Hybrid FRP is better than single FRP in suppressing the lateral expansion of the specimens.

Under axial loading, the specimens expanded. The expansion in lateral direction induces stresses in FRP. The amount of confining pressure of FRP depends on the stiffness and thickness of FRP. Therefore, degree of confinement provided by Hybrid FRP as well as stiffness of Hybrid FRP is double than that by single FRP. This leads to better performance of Hybrid FRP. Furthermore, this effect is more significant with thinner outer steel tube.

In addition, the combination of CFRP and GFRP as repair method is more superior to using multilayers of only one type of FRP. CFRP contributed to high tensile, high compressive strength, high stiffness and reduces the density; while GFRP possess high ultimate tensile strain and reduces the cost. Furthermore, using GFRP as first layer can prevent the long term effect of galvanic corrosion if outer steel tube was directly in contact with CFRP.

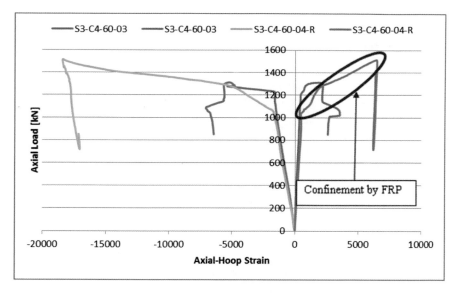

(a) 60 minutes

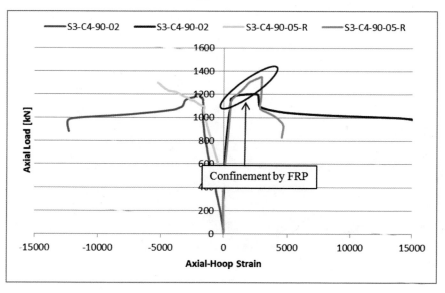

(b) 90 minutes

Fig. 5.17 Confinement effect of single layer CFRP for 60 and 90 min of fire exposure time

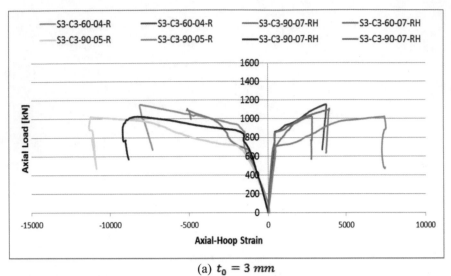

(a) $t_0 = 3\ mm$

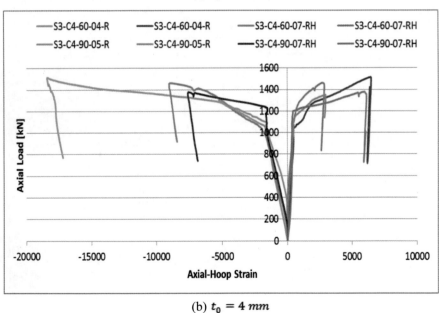

(b) $t_0 = 4\ mm$

Fig. 5.18 Axial load versus axial-hoop strain curves of single layer and Hybrid FRP

References

1. Tao, Z., Han, L.H., Wang, L.L.: Compressive and flexural behaviour of CFRP-repaired concrete-filled steel tubes after exposure to fire. J. Constr. Steel Res. **63**(8), 1116–1126 (2007)
2. Che, Y., Wang, Q.L., Shao, Y.B.: Compressive performances of the concrete filled circular CFRP-steel tube (C-CFRP-CFST). Adv. Steel Constr. **8**(4), 331–358 (2012)

Chapter 6
Conclusion and Recommendation

This research study is focused on three aspects of concrete-filled double skin steel tubular (CFDST) columns; (1) relationship between thickness of outer steel tube, t_0 and maximum temperature of concrete, (2) the residual strength of CFDST columns after fire exposure and (3) effectiveness of repair method using Single and Hybrid FRP on fire-damaged CFDST columns. Based on the results of experiment, the following conclusions can be made.

6.1 Relationship Between Thickness of Outer Steel Tube and Maximum Temperature of Concrete

This research study cleared contradictory finding by two previous researcher regarding the relationship between t_0 and maximum temperature of concrete. It is concluded that t_0 did not have significant effect on maximum temperature of concrete which seconded findings by Kodur [1].

6.2 Residual Strength of Concrete-Filled Double Skin Steel Tubular (CFDST) Columns After Fire Exposure

Residual strength of CFDST columns are classified using Residual Strength Index, RSI in this study. Apart from RSI, the behavior of heated CFDST columns is assessed based on secant stiffness and Ductility Index, DI. Therefore, the following conclusion can be made:

© The Author(s), under exclusive license to Springer Nature Singapore Pte Ltd. 2021
K. K. Choong et al., *Concrete-Filled Double Skin Steel Tubular Column with Hybrid Fibre Reinforced Polymer*, SpringerBriefs in Applied Sciences and Technology, https://doi.org/10.1007/978-981-16-2715-6_6

- RSI for $t_o = 4$ mm is lower than RSI for $t_o = 3$ mm regardless of fire exposure time. However, the performance of $t_o = 4$ mm is more significant when subjected to longer fire exposure time i.e., 90 min.
- Secant stiffness and DI increase as t_o increases. Specimens under 90 min of fire exposure time with $t_o = 4$ mm are found to become more ductile then the control specimens.

6.3   Effectiveness of Repair Method Using Single and Hybrid Fiber Reinforced Polymer (FRP) on Fire-Damaged Concrete-Filled Double Skin Steel Tubular (CFDST) Columns

The increment or reduction in strength is classified by calculating Strength Enhancement Index (SEI) of the repaired specimens. The performance of single and Hybrid FRP are assessed by comparing the SEI, secant stiffness and Ductility Index (DI) of the specimens. The increment in ultimate strength of Hybrid FRP is found to be superior to single FRP due to double confinement provided by FRP. Therefore, it can be concluded that Hybrid FRP is more effective than Single FRP in order to repair fire-damaged CFDST columns. In addition, the following conclusions can be drawn:

- Ultimate load of fire-damaged CFDST columns increased greatly after repairing with single layer of FRP. All specimens exposed to 60 min of fire exposure time (except S2-C4-60) experienced an increased in ultimate load exceeding the ultimate load of control specimen.
- SEI increased with decreased fire exposure time and decreased with increased t_o. FRP is found to be more effective in enhancing specimen with thinner t_o.
- Secant stiffness increased after the specimens are repaired with single layer of FRP. However, unlike ultimate load, the FRP wrap is unable to restore secant stiffness up to secant stiffness of control specimen.
- The DI of repaired specimens yielded mix results. In S1-C4-90 and S2-C4-60, the increment in DI nearly reached and exceeded its DI of control specimen, respectively. However, for other specimens, the increment is not up to the DI of control specimen.
- SEI, secant stiffness and DI increased with increase in number of FRP layer, i.e., from single layer to Hybrid FRP. However, similar to single FRP, the effectiveness of Hybrid FRP is more pronounced in the case of thinner t_o.

6.4   Recommendation for Future Work

This research study focuses on repairing of fire-damaged CFDST columns using Hybrid FRP. However, only two type of FRP used that make up the Hybrid FRP, namely CFRP and GFRP. In addition, the CFDST columns tested as an individual column not as a part of whole structure. Therefore, the following matters are recommended for further study:

- The performance of repaired fire-damaged CFDST columns with more Hybrid FRP repairing scheme since different FRP, i.e., Aramid FRP, Basalt FRP, possess different characteristic.
- Using Hybrid FRP to repair real fire-damaged structures where there is a need to consider columns as a part of whole structure.

Reference

1. Kodur, V.K.R.: Performance-based fire resistance design of concrete-filled steel columns. J. Constr. Steel Res. **51**(1), 21–36 (1999)

Printed in the United States
by Baker & Taylor Publisher Services